The History of Modern Physics

1800 – 1950

Volume 5

The History of Modern Physics, 1800–1950

The History of Modern Physics, 1800–1950

TITLES IN SERIES

INTRODUCTORY NOTE

The Tomash/American Institute of Physics series in the History of Modern Physics offers the opportunity to follow the evolution of physics from its classical period in the nineteenth century when it emerged as a distinct discipline, through the early decades of the twentieth century when its modern roots were established, into the middle years of this century when physicists continued to develop extraordinary theories and techniques. The one hundred and fifty years covered by the series, 1800 to 1950, were crucial to all mankind not only because profound evolutionary advances occurred but also because some of these led to such applications as the release of nuclear energy. Our primary intent has been to choose a collection of historically important literature which would make this most significant period readily accessible.

We believe that the history of physics is more than just the narrative of the development of theoretical concepts and experimental results: it is also about the physicists individually and as a group—how they pursued their separate tasks, their means of support and avenues of communication, and how they interacted with other elements of their contemporary society. To express these interwoven themes we have identified and selected four types of works: reprints of "classics" no longer readily available; original monographs and works of primary scholarship, some previously only privately circulated, which warrant wider distribution; anthologies of important articles here collected in one place; and dissertations, recently written, revised, and enhanced. Each book is prefaced by an introductory essay written by an acknowledged scholar, which, by placing the material in its historical context, makes the volume more valuable as a reference work.

The books in the series are all noteworthy additions to the literature of the history of physics. They have been selected for their merit, distinction, and uniqueness. We believe that they will be of interest not only to the advanced scholar in the history of physics, but to a much broader, less specialized group of readers who may wish to understand a science that has become a central force in society and an integral part of our twentieth-century culture. Taken in its entirety, the series will bring to the reader a comprehensive picture of this major discipline not readily achieved in any one work. Taken individually, the works selected will surely be enjoyed and valued in themselves.

The History of Modern Physics
1800 - 1950

Volume *5*

Physics for a New Century

Papers Presented at
the 1904 St. Louis Congress

A compilation selected
and a preface by
KATHERINE R. SOPKA

Introduction by
ALBERT E. MOYER

Tomash Publishers

American Institute of Physics

Library of Congress Cataloging in Publication Data
Main entry under title:

Physics for a new century.

(The History of modern physics, 1800–1950; v. 5)
   Bibliography: p.
   1. Physics—History—Congresses. I. Sopka, Katherine Russell. II.
Congress of Arts and Science (1904:   Saint Louis, Mo.) III. Series.
QC6.9.P48 1985   530'.09   85-28623
ISBN 0-88318-487-7

CONTENTS

Festival Hall, Central Cascade and Fountains

PREFACE

The papers reprinted in this volume were delivered during the week-long Congress of Arts and Science held in St. Louis in mid-September 1904 in conjunction with the Universal Exposition commemorating the centenary of the purchase by the United States of the Louisiana Territory. In the words of Howard J. Rogers, Director of Congresses and Chief of the Department of Education for the Exposition, "The exposition and the congress are correlative terms. The former concentres the visible products of the brain and the hand of man; the congress is the literary embodiment of its activities."

The notion of holding such a congress was not new but its conception at St. Louis in 1904 was. Groundwork for such events began with the establishment of an International Scientific Commission in connection with the Paris Exposition of 1867. Conferences on a variety of topics were held at subsequent expositions in Vienna in 1873, Philadelphia in 1876, and Paris in 1878. However, the idea of a fully developed congress associated with an exposition dated only to the one held in Paris in 1889 when some 70 conferences took place. Additional congresses convened during the World's Columbian Exposition in Chicago in 1893 and the Paris Exposition of 1900. Although numerous conferences were held during those expositions, they were frequently disjointed affairs with no overall plan, of interest to and attended by only a small circle of persons.

From the start, the Committee responsible for planning the St. Louis Congress was influenced by a plan set forth by Hugo Munsterberg, born and educated in Germany and then a Professor of Psychology at Harvard University. Munsterberg was convinced of the overall "Unity of Knowledge" and proposed that this Congress be organized around that central theme. In this way the more than 300 talks given could be put into mutual context and would correlate well with the theme of "Education" that had been adopted by the Exposition. Thus, in contrast to the disjointed conferences at previous expositions, the St. Louis Congress of Arts and Science would provide a unified view of mankind's intellectual achievements.

The word "Science" in the title meant, for Munsterberg and the Congress, all knowledge or learning rather than the more narrow connotation it has today. Science was thought of as an intellectual, theoretic human activity engaged in by individuals, complemented by "Arts" which, in turn, provided implementation beneficial to all. The program for the Congress would focus on the "Progress of Man" since the time of the Louisiana Purchase. It would include presentations in political history, economics, law, languages, fine arts, music, religion, metaphysics, mathematics, philosophy, physical and biological sciences, medicine, psychology, and technology. These were organized

ix

into Departments of "Sciences" designated Normative, Historical, Physical, Mental, Utilitarian, Regulative, and Cultural. Mathematics was placed, along with philosophy, in the Normative Department. Thus, the papers of Ludwig Boltzmann and Henri Poincaré were delivered there.

The Committee that Munsterberg persuaded to accept and implement his plan was an illustrious one which included Presidents Nicholas Murray Butler of Columbia University (Chairman), William R. Harper of the University of Chicago, R. H. Jesse of the University of Missouri, and Henry S. Pritchett of the Massachusetts Institute of Technology. Other members were Librarian of Congress Herbert Putnam, Director of the Field Columbian Museum Frederick J. V. Skiff, and Frederick W. Holls, a New York lawyer. As planning progressed, Simon Newcomb was designated President of the International Congress with Professors Munsterberg and Albion Small of the University of Chicago as Vice Presidents.

Planning had begun some two years in advance of the actual Congress. Once the overall plan was adopted and generous funding of $200,000 provided by the Executive Committee of the Exposition, the next step was the selection of speakers. To ensure a truly international character for the Congress a committee consisting of Newcomb, Munsterberg, and Small went to Europe in the Summer of 1903 to explain the plan and scope of the Congress and to deliver in person invitations to those scholars they hoped would participate. Each such individual would be given an honorarium in addition to having his travel expenses paid and being housed, fed, and entertained while in the United States. In all 150 invitations were extended, 117 acceptances secured.

Once the list of foreign invitees was established, American speakers for the Congress were also chosen on the advice of the various American professional societies. In addition, the Committee for the Congress, by its own account, tried to persuade members of those societies to attend the Congress by suggesting that they schedule their own annual meetings to be held in St. Louis either immediately before or just after the Congress. The American Physical Society did hold a meeting on the Friday before the Monday opening of the Congress. But the American Association for the Advancement of Science chose to meet in St. Louis the preceding December. In assessing the overall success of the Congress *Popular Science Monthly*, a widely read journal at that time, lamented the lack of coordination and participation with not only the AAAS but also the National Academy of Sciences, the Smithsonian Institution, and the Carnegie Institution. It also noted that "Officers of the army of science were paid to be present, but the rank and file of American workers were not there." (American speakers at the Congress also received travel expenses and honoraria.)

The 128 sessions which comprised the Congress were scheduled in buildings erected specifically for the Exposition and in some excellent new facilities of nearby Washington University. The opening exercises were held in the Festival Hall of the Exposition while the Administration Building of Washington University (see page 76) served as Headquarters for the Congress.

The week of the Congress contained social as well as intellectual activities. Not only were the full facilities of the Exposition available but evening festivals and receptions for participants were arranged. The foreign delegates were additionally entertained at evening banquets during the Congress and had been guests of the University of Chicago during the preceding week. At the close of the Congress they were transported by train to Washington, D.C., where they were given a reception by President Theodore Roosevelt. They subsequently were entertained as guests of a number of universities, including Harvard and Columbia, before departing the United States.

Extravagant encomiums were heaped on the Congress at its close. Speaking at one of the banquets, Director of the Division of Exhibits F. J. V. Skiff said:

"This congress is the peak of the mountain that this Exposition has builded on the highway of progress. From its heights we contemplate the past, record the present, and gaze into the future.

"This universal exposition is a world's university. The International Congress of Arts and Science constitutes the faculty; the materials on exhibition are the laboratories and the museums; the students are mankind."

Dr. Theodore Escherich of the University of Vienna responded:

"The conception of this International Congress of all Sciences in its originality and audacity, in its universality and comprehensive organization, is truly a child of the 'young-American spirit'...

"After this Congress has come to a close and the collection of the lectures delivered, an unparalleled encyclopaedia of human knowledge, both in extent and content, will have appeared. We may say that this Fair has become of epochal importance, not alone for trade and manufactures, but also for science. These proud palaces will long have disappeared and been forgotten when this work, a *monumentum aere perennius*, shall still testify to future generations the standard of scientific attainment at the beginning of the twentieth century."

Hugo Munsterberg, apparently well satisfied with the success of this plan, viewed the Congress as "the first academic alliance between the United States and Europe;...the first great undertaking in which the Old and the New Worlds stood on equal levels and in which Europe really became acquainted with the scientific life of the United States."

From the start, it was intended that the Proceedings of the Congress would be published. Thus, according to Howard J. Rogers, they would "stand as a monument to the breadth and enterprise of the Exposition long after its buildings had disappeared and its commercial achievements grown dim in the minds of men." Eight volumes were so issued in 1906 by Houghton Mifflin and Co. In a sense, by extracting and reprinting now only those papers related to physics, we are violating the intention of the organizers to display all knowledge. However, as will be pointed out by Albert Moyer, physics in 1904 was at a critical stage of development and we believe that historians of science today will welcome the easy access which this volume provides to the perspective on contemporary physics provided by the illustrious group of authors represented.

Acknowledgments

The information summarized above was obtained principally from "The History of the Congress" by Howard J. Rogers and "The Scientific Plan of the Congress" by Hugo Munsterberg, both of which are included in Volume I of the Proceedings. In addition, I found useful an article by Munsterberg in *The Atlantic Monthly* of May 1903 and several commentaries on Congress activities published in *Popular Science Monthly*.

Katherine R. Sopka

September 1985

FOREWORD

by
Albert E. Moyer

The 1904 Congress of Arts and Science occurred during a period of intense experimental, theoretical, and ideological ferment in physics. Expectations of change existed even though the physics community was a year away from learning about Albert Einstein's theory of relativity and several years away from fully acknowledging Max Planck's recent quantum hypothesis.

Though the ferment was rooted in events dating back many decades, Wilhelm Röntgen's discovery of x rays in late 1895 signaled a quickening of activity. Röntgen's x rays, accidentally detected, and having unanticipated penetrative properties, caused a sensation among scientists. During the following year, Henri Becquerel made an equally fortuitous and unsettling discovery, radioactivity. This in turn quickly led to evidence for atomic transmutation with the work of Marie and Pierre Curie, Ernest Rutherford, and Frederick Soddy. Meanwhile, J. J. Thomson's experimental characterization of the electron in 1897 compounded the general excitement of this period.

But physicists around 1904 perceived this era to be more than one of novel experimental results. They were aware of associated unrest that was occurring in the realm of theory. Specifically, the newer statistical-mechanical, thermodynamic, and electromagnetic views of physical phenomena were in vibrant interplay with each other and with the more traditional mechanical outlook— the outlook that all physical phenomena could be represented in terms of underlying atoms, molecules, and ethers that obey the laws of classical mechanics. Although many scientists still espoused this traditional program of mechanical explanation, an increasing number of researchers were challenging and criticizing it, often on explicitly operational or positivistic grounds. Moreover, certain mathematically adept physicists hoped to subordinate orthodox mechanics to generalized statistical mechanics. In addition, scientists with a more phenomenological perspective, as advocated by Wilhelm Ostwald, aimed to install thermodynamics as the cornerstone of physical explanation. Others sought an alternative to orthodox mechanism in electromagnetic theory, particularly as presented by H. A. Lorentz. Scientists holding the extreme of this electromagnetic viewpoint hoped to reduce all of classical mechanics to electromagnetic concepts.

Finally in the period around 1904, there also was unrest in the realm of scientific ideology. Physicists were increasingly abandoning their earlier belief in an omniscient science that steadily progresses toward orderly and absolute natural laws. More and more, they were resigning themselves to metaphysically neutral descriptions or correlations of phenomena. Victorian confidence was giving way to sober scientific skepticism.[1]

This conceptual ferment—experimental, theoretical, and ideological—was dramatized in 1904 when many of the world's principal physicists assembled at the Congress of Arts and Science. Journeying to St. Louis was a group of international leaders whose names today connote scientific upheaval: Henri Poincaré, Paul Langevin, Ernest Rutherford, Wilhelm Ostwald, and Ludwig Boltzmann. These men shared the St. Louis podium with a similarly progressive but less well known group of Americans including Carl Barus, Arthur Kimball, Dewitt Brace, Edward Nichols, and Robert Millikan.[2]

Foreign Delegates

University of Paris professor Henri Poincaré (1854–1912) set the tone of the Congress in his speech on "mathematical physics"—a speech wherein he sought to evaluate contemporary physics by placing it within a historical context. In Poincaré's view, the Newtonian physics of detailed "central forces" acting between atoms had given way during the nineteenth century to the physics of general "principles," such as the principle of the conservation of energy. He speculated, moreover, that the physics of "principles" was currently entering a stage of "crisis" and yielding to a still veiled third phase of physics. The second law of thermodynamics, for example, was in question; evidence suggested that it was merely an "imperfect" statistical theorem concerning molecular motions. The principle of the conservation of mass was also in doubt. Indeed, some physicists were proposing that there was no "true mechanical mass" but only mass of "exclusively electrodynamic origin." Many more physicists were affirming that mass, whether mechanical or electrodynamic, varied with speed. These challenges to traditional thought suggested to Poincaré possible forms of the emerging, third phase of physics. Perhaps statistical mechanics "is about to undergo development and serve as model" for all of physics. Alternatively, perhaps there will soon emerge "a whole new mechanics...where inertia increasing with the velocity, the velocity of light would become an impassable limit."[3] With this latter prediction, along with an equally incisive evaluation of the principle of relativity, Poincaré was approaching in significant part the breakthrough that Einstein actually achieved the following year.[4]

Besides surveying recent shifts in the substantive content of physics, Poincaré also traced the development of three distinct views of natural law. Prior to the adoption of the physics of central forces, scientists considered a law of nature to be "an internal harmony, static, so to say, and immutable." With the advent of the physics of central forces, and the example of Newton's law of gravitation, scientists gradually discarded this "ancient" teleological view.

Natural law became—and remained into Poincaré's second and current phase of history, that of principles—simply a philosophically neutral, "constant relation between the phenomenon of to-day and that of to-morrow." Finally, envisioning a future physics based on statistical mechanics, Poincaré hazarded a guess on the coming attitude: "Physical law...will take the character of a statistical law." For Poincaré in 1904, the ideology as well as the substance of nineteenth-century physics was exceedingly vulnerable to a "profound transformation."[5]

Equally convinced as Poincaré of fundamental changes occurring in physics was Paul Langevin (1872–1946), a young professor of physics at the Collège de France. He felt that the scientific synthesis promised by the physics of the electron, his St. Louis topic, was like "a kind of New America...which can teach many things to the Old World." Confident of the "electronic conception of matter," Langevin advocated a complete inversion of traditional scientific thought. No longer should scientists seek a mechanical representation of electromagnetic concepts, but rather "an electromagnetic representation of the principles and ideas of ordinary, material mechanics." And for Langevin in 1904, this electromagnetic view was more than merely a possibility. In concluding his speech, he emphasized that the electromagnetic revolution was well under way.[6]

In a lengthy but tightly knit talk, Ernest Rutherford (1871–1937), then a professor at McGill University in Montreal, appraised the latest experimental and theoretical knowledge regarding all phases of radioactivity—from the nature of alpha, beta, and gamma emanations to the source of energy of radioactive bodies. Rutherford was especially concerned to show the amenability of radioactivity to Langevin's type of electromagnetic interpretation. He was sympathetic to the idea that the mass of a beta particle, or electron, was entirely of electromagnetic origin and that the mass increased with speed up to the limiting speed of light. Moreover, he carefully pointed out that his and Frederick Soddy's radioactive disintegration theory, regarding questions of energy conservation, was in harmony with "the modern views of the electronic constitution of matter."[7]

Incidentally, although Rutherford came from McGill University in Canada and was officially a "foreign delegate" to the St. Louis Congress, he had become active in the physics community in the United States. Thus, it is not surprising that Robert Millikan (1868–1953), an aspiring thirty-six-year-old professor at the University of Chicago, chose as the topic for a short paper at the St. Louis Congress, "The Relation Between the Radioactivity and the Uranium Content of Certain Minerals"—a topic popularized by Rutherford's main American collaborator, Bertram Boltwood of Yale.[8]

Two delegates to the Congress, participants in the section on "Methodology of Science," explicitly addressed philosophical problems pertinent to physics. Benno Erdmann (1851–1921), a philosopher from the University of Bonn, reevaluated traditional conceptions of causal law. His goal was to move beyond either an exclusively empiricist or rationalist account of causality as advocated by most thinkers from the time of Hume and Kant up through Ernst Mach. Keeping in mind the views of recent scientists such as Hermann von Helmholtz and Heinrich Hertz, Erdmann arrived at a rather eclectic but essentially classical notion of causality. Specifically, he defended the idea of causal law by proposing "a universal phenomenological dynamism" in which there exists a "dynamic intermediary between cause and effect."[9]

The featured speaker on methodology of science was not a philosopher but Wilhelm Ostwald (1853–1932), professor of physical chemistry at Leipzig. Ostwald stressed the importance of establishing a definite and complete "correspondence" between the actual "manifold" of experience in a certain empirical situation and the formal "manifold" of the scientists' conceptual constructions whether they be words, equations, or symbols. Regarding physical and chemical phenomena, he specifically insisted that the science of energetics, with its emphasis on measurable energy transformations, promised the fullest possible agreement between the empirical and conceptual manifolds. Indeed, Ostwald implied that in both physics and chemistry the newer, phenomenological, "energetical" outlook—an outlook that he had pioneered—was superior to the older, hypothetical, mechanical and atomistic view. Judged by the criterion of direct correspondence between concepts and empirical data, "all previous systematizations in the form of hypotheses" were deficient.[10]

Not everyone in Ostwald's St. Louis audience agreed with the energeticist's contentions. Robert Millikan afterward remembered that the "question of whether the atomic and kinetic theories were essential" was the Congress's "chief subject of debate." In fact, sitting in the front row for Ostwald's lecture and joining in the following discussion was one of the oldest and most distinguished of the foreign representatives to the Congress, Ludwig Boltzmann (1840–1906), Ostwald's scientific adversary from Leipzig and, more recently, Vienna.[11] On the day after Ostwald's lecture, Boltzmann made a formal rebuttal in his own talk on theoretical physics delivered before the section on "Applied Mathematics." While granting that physics was presently in disarray—perhaps even "in process of revolution"—Boltzmann did not agree that generalized, phenomenological theories were better equipped to reestablish order than specialized, mechanical hypotheses involving atoms, molecules, and ether. Though appreciative of phenomenological theories with their stress on direct observation, he denied that such theories were free from hypotheses or idealizations and hence irrefutable. This included "the so-called theory of energy." Moreover, Boltzmann felt that the usefulness of phenomenological theories was limited merely to summarizing or developing "knowledge previously acquired." On the other hand, specialized and admittedly tentative hypotheses "give the imagination room for play" and thus inspire "the most unexpected discoveries." He also felt that the recent web of experiments involving cathode rays and radioactivity added credence to the atomistic viewpoint. Boltzmann was particularly optimistic about statistical mechanics, especially J. Willard Gibbs' formulation.[12]

American Delegates

A number of the American physicists at the St. Louis Congress joined with Poincaré, Langevin, Rutherford, Ostwald, and Boltzmann in saying that physical science was undergoing a dramatic upheaval. Though less personally involved in this restructuring of science than their foreign colleagues, the Americans were well informed about the latest electromagnetic, thermodynamic, and statistical mechanical alternatives to the faltering classical program of mechanical explanation. And, though less articulate than their overseas visitors, they generally advocated the newer phenomenological, operational, or positivistic attitudes toward research and physical law. These more progressive members of the American delegation included: Carl Barus (1856–1935), physicist at Brown University with an earlier background in geophysics; Arthur Kimball (1856–1922), professor at Amherst College; Dewitt Brace (1859–1905), University of Nebraska researcher with a special interest in the ether; and, as mentioned before, Robert Millikan (1868–1953), physicist at the University of Chicago. A more moderate member of this group was Edward Nichols (1854–1937), Cornell professor and founder in 1893 of *The Physical Review.*

Of course, not all the Americans who addressed the Congress were comfortable and conversant with the latest trends. The president of the Congress and keynote speaker was sixty-nine-year-old Simon Newcomb (1835–1909), a venerated astronomer, mathematician, and physicist most at home with the science of the nineteenth-century.[13] Similarly, Robert Woodward (1849–1924), president of the newly founded Carnegie Institution of Washington, opened the meetings of the Division of Physical Science with remarks true to the scientific ethos of the recent past. In words reminiscent of former decades, he characterized all physical phenomena as being atomic or molecular in nature and he hoped for a grand mechanical unification of these phenomena: "the day seems not far distant when there will be room for a new *Principia* and for a treatise which will accomplish for molecular systems what the *Mécanique Céleste* accomplished for the solar system."[14]

Francis Nipher (1847–1926), professor of physics at Washington University in St. Louis, also spoke out at the Congress in favor of the threatened mechanistic program. His reaction to the turn-of-the-century proliferation of experimental and theoretical novelties was retrenchment. He sought to subsume new phenomena and concepts such as radioactivity and the electron under established mechanical patterns. For Nipher, radioactivity was not necessarily an unprecedented phenomenon. It was possibly a type of "explosion" that happened to display normal explosive properties "to a very exalted degree." In like manner, he objected to replacing the seasoned mechanical view of nature with the electromagnetic view.[15]

Regardless of the traditional leanings of Nipher, Woodward, and Newcomb, the new electromagnetic view was the dominant outlook among the more progressive Americans in St. Louis. Carl Barus capped his encyclopedic

xvii

survey of "The Progress of Physics in the Nineteenth Century" by describing the recent and "splendid triumph of the electronic theory."[16] Though admitting to having been "startled at first by the very audacity of this theory," Arthur Kimball echoed Barus's sentiment. Kimball was particularly impressed by J. J. Thomson's "most remarkable" theory of the atom as a swirl of hundreds of electrons. In general, he concurred that physicists should "seek the explanation of matter and its laws in terms of the properties of ether and electricity, instead of trying to unravel the secrets of electricity and ether in terms of matter and motion."[17]

Dewitt Brace heeded this advice. An international authority on the ether, Brace reported to the Congress that he was experimentally testing the implications of electron theory regarding, for example, mass variation and length contraction.[18] Finally, while Edward Nichols accepted the unprecedented empirical effects associated with electron theory, he preferred to interpret the effects in terms of an underlying, abstract ether rather than in terms of primitive, "disembodied" electric charge. That is, in the increasingly common refrain, "matter is composed of electricity and of nothing else," he preferred to substitute for the hypothetical concept of electricity the equally "imaginary" but more tractable concept of ether. "If matter is to be regarded as a product of certain operations performed upon the ether," he reasoned, "there is no theoretical difficulty about transmutation of elements, variation of mass, or even the complete disappearance or creation of matter."[19]

These Americans, however, were not naive champions of the electron theory and related views. They realized along with their European colleagues that they were dealing with a nascent theory, incomplete and unproven. Indeed, so penetrating were some of the criticisms that physicists would resolve them only much later, after accepting and developing Planck's quantum hypothesis and Einstein's special and general theories of relativity. Kimball wondered, for example, if the overall theory would ever incorporate gravitation. Moreover, regarding a specific branch of electron theory, he pinpointed the tasks still to be accomplished by J. J. Thomson's electron model of the atom. Kimball could only hope that this model would eventually clarify the seemingly anomalous behavior of specific heats of gases. Similarly, he could merely trust that Thomson's idea would someday account for complex atomic spectra. And he worried that the model still lacked an agency for assigning distinct numbers of electrons to the atoms of particular chemical elements. "Some kind of natural selection seems to be needed," he wrote, "to explain why some atoms having special numbers of corpuscles survive while intermediate ones are eliminated."[20] Brace also had reservations about the electron theory. Specifically, he was dissatisfied with the seemingly *ad hoc* character of recent electromagnetic interpretations of ether-drift experiments—interpretations that appeared "highly artificial" in their "successive auxiliary hypotheses and approximations."[21]

The Americans at the Congress coupled their conditional endorsement of the electron theory with an open wariness of prior mechanistic outlooks.

xviii

Barus stressed that while all physicists now accepted Maxwell's electromagnetic equations as accurate descriptions of phenomena, most rejected the mechanical "methods by which Maxwell arrived at his great discoveries." Along the same line, in a paraphrase of Poincaré, Barus emphasized the arbitrariness of all interpretations of electromagnetism that involved an "ether mechanism."[22] Kimball likewise called attention to the hypothetical and heuristic nature of the mechanical postulates traditionally associated with theories of matter, such as the vortex-atom theory.[23]

Edward Nichols, while more tied than Barus and Kimball to traditional mechanical views, also displayed a discriminating attitude toward these views. Specifically, Nichols defended the increasingly familiar idea of evaluating scientific concepts through use of the "dimensional formula." His premise was that any physical quantity should ultimately be expressible as a relationship between three empirically certain, mechanical concepts—mass, distance, and time. This mode of analysis, with its requirement of precise definitions built on direct observations, afforded "a valuable criterion of the extent and boundaries of our strictly definite knowledge of physics." In general, like positivists such as Ernst Mach, Nichols was content to view science as "nothing more than *an attempt to classify and correlate our sensations.*"[24]

In their speeches, the foreign and American delegates to the St. Louis congress highlighted the fundamental tensions pervading turn-of-the-century physics. Whereas many of the experimental, theoretical, and ideological precepts they voiced would soon be forgotten, others would be incorporated in the new physics of relativity and the quantum. The Congress of 1904 thus affords a rare and uniquely condensed glimpse of physics in transition.

1. P. M. Harman, *Energy, Force, and Matter: The Conceptual Development of Nineteenth-Century Physics* (Cambridge: Cambridge Univ. Press, 1982). Russell McCormmach, "H. A. Lorentz and the Electromagnetic View of Nature," Isis **61** (1970), 459–97. See also Daniel J. Kevles, *The Physicists: The History of a Scientific Community in Modern America* (New York: Knopf, 1978), pp. 3–90.

2. For a slightly more detailed discussion of the Congress, see Albert E. Moyer, *American Physics in Transition: A History of Conceptual Change in the Late Nineteenth Century* (Los Angeles: Tomash, 1983), pp. 142–61. See also Erwin N. Hiebert, "The State of Physics at the Turn of the Century," in *Rutherford and Physics at the Turn of the Century*, edited by Mario A. Bunge and William R. Shea (New York: Watson, 1979), pp. 1–28.

3. Poincaré, "The Principles of Mathematical Physics," translated by George Halsted, in *Philosophy and Mathematics*, Vol. I of *Congress of Arts and Science: Universal Exposition, St. Louis, 1904*, edited by Howard Rogers (Boston: Houghton Mifflin, 1905), pp. 606–80, 609–10, 615, 621. Hereafter cited as *Congress I*.

4. Poincaré, pp. 610–12. For a discussion, see Jeremy Bernstein, *Einstein* (New York: Viking Press, 1973), pp. 86–87; Gerald Holton, "Poincaré and Relativity," in *Thematic Origins of Scientific Thought: Kepler to Einstein* (Cambridge: Harvard Univ. Press, 1973), pp. 185–195; and Loyd Swenson, *Ethereal Aether: A History of the Michelson-Morley-Miller Aether-Drift Experiments, 1880–1930* (Austin: Univ. of Texas Press, 1972), pp. 149–51.

5. Poincaré, pp. 604–06, 621.

6. Langevin, "The Relations of Physics of Electrons to Other Branches of Science," translated by Bergen Davis, in *Inorganic Sciences*, Vol. IV of *Congress of Arts and Science*, pp. 121, 138, 156. Hereafter cited as *Congress IV*.

7. Rutherford, "Present Problems of Radioactivity," in *Congress IV*, pp. 158–159, 171.

8. Millikan, "Abstract of 'The Relation Between the Radioactivity and the Uranium Content of Certain Minerals'," in *Congress IV*, p. 187.

9. Erdmann, "The Content and Validity of the Causal Law," translated by Walter T. Marvin, in *Congress I*, pp. 373–74, 379–88.

10. Ostwald, "On the Theory of Science," translated by R. M. Yerkes, in *Congress I*, pp. 347–49.

11. William H. Davis, "The International Congress of Arts and Science," Pop. Sci. Mo. **66** (1904). Robert A. Millikan, *Autobiography* (New York: Prentice-Hall, 1950), pp. 22, 84–85.

12. Boltzmann, "The Relations of Applied Mathematics," translated by S. Epsteen, in *Congress I*, pp. 592–95, 599, 601–03.

13. Newcomb, "The Evolution of the Scientific Investigator," in *Congress I*, pp. 135–47.

14. Woodward, "The Unity of Physical Science," *Congress IV*, p. 5.

15. Nipher, "Present Problems in the Physics of Matter," *Congress IV*, pp. 87–88, 92, 98–101.

16. Barus, "The Progress of Physics in the Nineteenth Century," *Congress IV*, pp. 64–65.

17. Kimball, "The Relations of the Science of Physics of Matter to Other Branches of Learning," *Congress IV*, pp. 74–76.

18. Brace, "The Ether and Moving Matter, " *Congress IV*, pp. 113–117.

19. Nichols, "The Fundamental Concepts of Physical Science," *Congress IV*, pp. 24–27.

20. Kimball, pp. 74–76.

21. Brace, pp. 116–117.

22. Barus, pp. 62–64.

23. Kimball, pp. 72–73.

24. Nichols, pp. 18–21, 27.

CONGRESS OF ARTS AND SCIENCE

UNIVERSAL EXPOSITION, ST. LOUIS, 1904

EDITED BY

HOWARD J. ROGERS, A.M., LL.D.

DIRECTOR OF CONGRESSES

VOLUME I

HISTORY OF THE CONGRESS
By THE EDITOR

SCIENTIFIC PLAN OF THE CONGRESS
By PROFESSOR HUGO MÜNSTERBERG

PHILOSOPHY AND MATHEMATICS

BOSTON AND NEW YORK
HOUGHTON, MIFFLIN AND COMPANY
The Riverside Press, Cambridge
1905

Simon Newcomb
1835–1909

PROCEEDINGS OF THE CONGRESS

INTRODUCTORY ADDRESS

DELIVERED AT THE OPENING EXERCISES AT FESTIVAL HALL BY
PROFESSOR SIMON NEWCOMB, PRESIDENT OF THE CONGRESS

THE EVOLUTION OF THE SCIENTIFIC INVESTIGATOR

As we look at the assemblage gathered in this hall, comprising so many names of widest renown in every branch of learning,—we might almost say in every field of human endeavor,—the first inquiry suggested must be after the object of our meeting. The answer is, that our purpose corresponds to the eminence of the assemblage. We aim at nothing less than a survey of the realm of knowledge, as comprehensive as is permitted by the limitations of time and space. The organizers of our Congress have honored me with the charge of presenting such preliminary view of its field as may make clear the spirit of our undertaking.

Certain tendencies characteristic of the science of our day clearly suggest the direction of our thoughts most appropriate to the occasion. Among the strongest of these is one toward laying greater stress on questions of the beginning of things, and regarding a knowledge of the laws of development of any object of study as necessary to the understanding of its present form. It may be conceded that the principle here involved is as applicable in the broad field before us as in a special research into the properties of the minutest organism. It therefore seems meet that we should begin by inquiring what agency has brought about the remarkable development of science to which the world of to-day bears witness. This view is recognized in the plan of our proceedings, by providing for each great department of knowledge a review of its progress during the century that has elapsed since the great event commemorated by the scenes outside this hall. But such reviews do not make up that general survey of science at large which is necessary to the development of our theme, and which must include the action of causes that had their origin long before our time. The movement which culminated in making the nineteenth century ever memorable in history is the outcome of a long series of causes, acting through many centuries, which are worthy of especial attention on such an occasion as this. In setting them forth we should avoid laying stress on those visible manifestations which, striking the eye of every beholder, are in no danger of being overlooked, and search rather for those agencies whose activities underlie the whole visible scene, but which are liable to be blotted out of sight by the very brilliancy of the results to which they have given rise. It is easy to draw attention to the wonderful qualities of the oak; but from that very fact, it may be needful to point out that the real wonder lies concealed in the acorn from which it grew.

Our inquiry into the logical order of the causes which have made our civilization what it is to-day will be facilitated by bringing to mind certain elementary considerations — ideas so familiar that setting them forth may seem like citing a body of truisms — and yet so frequently overlooked, not only individually, but in their relation to each other, that the conclusion to which they lead may be lost to sight. One of these propositions is that psychical rather than material causes are those which we should regard as fundamental in directing the development of the social organism. The human intellect is the really active agent in every branch of endeavor, — the *primum mobile* of civilization, — and all those material manifestations to which our attention is so often directed are to be regarded as secondary to this first agency. If it be true that "in the world is nothing great but man; in man is nothing great but mind," then should the keynote of our discourse be the recognition of this first and greatest of powers.

Another well-known fact is that those applications of the forces of nature to the promotion of human welfare which have made our age what it is, are of such comparatively recent origin that we need go back only a single century to antedate their most important features, and scarcely more than four centuries to find their beginning. It follows that the subject of our inquiry should be the commencement, not many centuries ago, of a certain new form of intellectual activity.

Having gained this point of view, our next inquiry will be into the nature of that activity, and its relation to the stages of progress which preceded and followed its beginning. The superficial observer, who sees the oak but forgets the acorn, might tell us that the special qualities which have brought out such great results are expert scientific knowledge and rare ingenuity, directed to the application of the powers of steam and electricity. From this point of view the great inventors and the great captains of industry were the first agents in bringing about the modern era. But the more careful inquirer will see that the work of these men was possible only through

4

a knowledge of the laws of nature, which had been gained by men whose work took precedence of theirs in logical order, and that success in invention has been measured by completeness in such knowledge. While giving all due honor to the great inventors, let us remember that the first place is that of the great investigators, whose forceful intellects opened the way to secrets previously hidden from men. Let it be an honor and not a reproach to these men, that they were not actuated by the love of gain, and did not keep utilitarian ends in view in the pursuit of their researches. If it seems that in neglecting such ends they were leaving undone the most important part of their work, let us remember that nature turns a forbidding face to those who pay her court with the hope of gain, and is responsive only to those suitors whose love for her is pure and undefiled. Not only is the special genius required in the investigator not that generally best adapted to applying the discoveries which he makes, but the result of his having sordid ends in view would be to narrow the field of his efforts, and exercise a depressing effect upon his activities. The true man of science has no such expression in his vocabulary as "useful knowledge." His domain is as wide as nature itself, and he best fulfills his mission when he leaves to others the task of applying the knowledge he gives to the world.

We have here the explanation of the well-known fact that the functions of the investigator of the laws of nature, and of the inventor who applies these laws to utilitarian purposes, are rarely united in the same person. If the one conspicuous exception which the past century presents to this rule is not unique, we should probably have to go back to Watt to find another.

From this viewpoint it is clear that the primary agent in the movement which has elevated man to the masterful position he now occupies, is the scientific investigator. He it is whose work has deprived plague and pestilence of their terrors, alleviated human suffering, girdled the earth with the electric wire, bound the continent with the iron way, and made neighbors of the most distant nations. As the first agent which has made possible this meeting of his representatives, let his evolution be this day our worthy theme. As we follow the evolution of an organism by studying the stages of its growth, so we have to show how the work of the scientific investigator is related to the ineffectual efforts of his predecessors.

In our time we think of the process of development in nature as one going continuously forward through the combination of the opposite processes of evolution and dissolution. The tendency of our thought has been in the direction of banishing cataclysms to the theological limbo, and viewing nature as a sleepless plodder, endowed with infinite patience, waiting through long ages for results. I do not contest the truth of the principle of continuity on which

5

this view is based. But it fails to make known to us the whole truth. The building of a ship from the time that her keel is laid until she is making her way across the ocean is a slow and gradual process; yet there is a cataclysmic epoch opening up a new era in her history. It is the moment when, after lying for months or years a dead, inert, immovable mass, she is suddenly endowed with the power of motion, and, as if imbued with life, glides into the stream, eager to begin the career for which she was designed.

I think it is thus in the development of humanity. Long ages may pass during which a race, to all external observation, appears to be making no real progress. Additions may be made to learning, and the records of history may constantly grow, but there is nothing in its sphere of thought, or in the features of its life, that can be called essentially new. Yet, nature may have been all along slowly working in a way which evades our scrutiny until the result of her operations suddenly appears in a new and revolutionary movement, carrying the race to a higher plane of civilization.

It is not difficult to point out such epochs in human progress. The greatest of all, because it was the first, is one of which we find no record either in written or geological history. It was the epoch when our progenitors first took conscious thought of the morrow, first used the crude weapons which nature had placed within their reach to kill their prey, first built a fire to warm their bodies and cook their food. I love to fancy that there was some one first man, the Adam of evolution, who did all this, and who used the power thus acquired to show his fellows how they might profit by his example. When the members of the tribe or community which he gathered around him began to conceive of life as a whole, — to include yesterday, to-day, and to-morrow in the same mental grasp — to think how they might apply the gifts of nature to their own uses, — a movement was begun which should ultimately lead to civilization.

Long indeed must have been the ages required for the development of this rudest primitive community into the civilization revealed to us by the most ancient tablets of Egypt and Assyria. After spoken language was developed, and after the rude representation of ideas by visible marks drawn to resemble them had long been practiced, some Cadmus must have invented an alphabet. When the use of written language was thus introduced, the word of command ceased to be confined to the range of the human voice, and it became possible for master minds to extend their influence as far as a written message could be carried. Then were communities gathered into provinces; provinces into kingdoms; kingdoms into the great empires of antiquity. Then arose a stage of civilization which we find pictured in the most ancient records, — a stage in which men were governed by laws that were perhaps as wisely adapted to their conditions as our laws are to ours,—in which the phenomena of nature were rudely observed, and striking occurrences in the earth or in the heavens recorded in the annals of the nation.

Vast was the progress of knowledge during the interval between these empires and the century in which modern science began. Yet, if I am right in making a distinction between the slow and regular steps of progress, each growing naturally out of that which preceded it, and the entrance of the mind at some fairly definite epoch into an entirely new sphere of activity, it would appear that there was only one such epoch during the entire interval. This was when abstract geometrical reasoning commenced, and astronomical observations aiming at precision were recorded, compared, and discussed. Closely associated with it must have been the construction of the forms of logic. The radical difference between the demonstration of a theorem of geometry and the reasoning of every-day life which the masses of men must have practiced from the beginning, and which few even to-day ever get beyond, is so evident at a glance that I need not dwell upon it. The principal feature of this advance is that, by one of those antinomies of the human intellect of which examples are not wanting even in our own time, the development of abstract ideas preceded the concrete knowledge of natural phenomena. When we reflect that in the geometry of Euclid the science of space was brought to such logical perfection that even to-day its teachers are not agreed as to the practicability of any great improvement upon it, we cannot avoid the feeling that a very slight change in the direction of the intellectual activity of the Greeks would have led to the beginning of natural science. But it would seem that the very purity and perfection which was aimed at in their system of geometry stood in the way of any extension or application of its methods and spirit to the field of nature. One example of this is worthy of atten- tion. In modern teaching the idea of magnitude as generated by motion is freely introduced. A line is described by a moving point; a plane by a moving line; a solid by a moving plane. It may, at first sight, seem singular that this conception finds no place in the Euclid- ian system. But we may regard the omission as a mark of logical purity and rigor. Had the real or supposed advantages of introduc- ing motion into geometrical conceptions been suggested to Euclid, we may suppose him to have replied that the theorems of space are independent of time; that the idea of motion necessarily implies time, and that, in consequence, to avail ourselves of it would be to introduce an extraneous element into geometry.

It is quite possible that the contempt of the ancient philosophers for the practical application of their science, which has continued in some form to our own time, and which is not altogether unwholesome, was a powerful factor in the same direction. The result was that, in keeping geometry pure from ideas which did not belong to it, it failed to form what might otherwise have been the basis of physical science. Its founders missed the discovery that methods similar to those of geometric demonstration could be extended into other and wider fields than that of space. Thus not only the development of applied geometry, but the reduction of other conceptions to a rigorous mathematical form was indefinitely postponed.

Astronomy is necessarily a science of observation pure and simple, in which experiment can have no place except as an auxiliary. The vague accounts of striking celestial phenomena handed down by the priests and astrologers of antiquity were followed in the time of the Greeks by observations having, in form at least, a rude approach to precision, though nothing like the degree of precision that the astronomer of to-day would reach with the naked eye, aided by such instruments as he could fashion from the tools at the command of the ancients.

The rude observations commenced by the Babylonians were continued with gradually improving instruments, — first by the Greeks and afterward by the Arabs, — but the results failed to afford any insight into the true relation of the earth to the heavens. What was most remarkable in this failure is that, to take a first step forward which would have led on to success, no more was necessary than a course of abstract thinking vastly easier than that required for working out the problems of geometry. That space is infinite is an unexpressed axiom, tacitly assumed by Euclid and his successors. Combining this with the most elementary consideration of the properties of the triangle, it would be seen that a body of any given size could be placed at such a distance in space as to appear to us like a point. Hence a body as large as our earth, which was known to be a globe from the time that the ancient Phœnicians navigated the Mediterranean, if placed in the heavens at a sufficient distance, would look like a star. The obvious conclusion that the stars might be bodies like our globe, shining either by their own light or by that of the sun, would have been a first step to the understanding of the true system of the world.

There is historic evidence that this deduction did not wholly escape the Greek thinkers. It is true that the critical student will assign little weight to the current belief that the vague theory of Pythagoras — that fire was at the centre of all things — implies a conception of the heliocentric theory of the solar system. But the testimony of Archimedes, confused though it is in form, leaves no serious doubt that Aristarchus of Samos not only propounded the view that the earth revolves both on its own axis and around the sun, but that he correctly removed the great stumbling-block in the way of this theory by adding that the distance of the fixed stars was

8

infinitely greater than the dimensions of the earth's orbit. Even the world of philosophy was not yet ready for this conception, and, so far from seeing the reasonableness of the explanation, we find Ptolemy arguing against the rotation of the earth on grounds which careful observations of the phenomena around him would have shown to be ill-founded.

Physical science, if we can apply that term to an uncoördinated body of facts, was successfully cultivated from the earliest times. Something must have been known of the properties of metals, and the art of extracting them from their ores must have been practiced, from the time that coins and medals were first stamped. The properties of the most common compounds were discovered by alchemists in their vain search for the philosopher's stone, but no actual progress worthy of the name rewarded the practitioners of the black art.

Perhaps the first approach to a correct method was that of Archimedes, who by much thinking worked out the law of the lever, reached the conception of the centre of gravity, and demonstrated the first principles of hydrostatics. It is remarkable that he did not extend his researches into the phenomena of motion, whether spontaneous or produced by force. The stationary condition of the human intellect is most strikingly illustrated by the fact that not until the time of Leonardo was any substantial advance made on his discovery. To sum up in one sentence the most characteristic feature of ancient and medieval science, we see a notable contrast between the precision of thought implied in the construction and demonstration of geometrical theorems and the vague indefinite character of the ideas of natural phenomena generally, a contrast which did not disappear until the foundations of modern science began to be laid.

We should miss the most essential point of the difference between medieval and modern learning if we looked upon it as mainly a difference either in the precision or the amount of knowledge. The development of both of these qualities would, under any circumstances, have been slow and gradual, but sure. We can hardly suppose that any one generation, or even any one century, would have seen the complete substitution of exact for inexact ideas. Slowness of growth is as inevitable in the case of knowledge as in that of a growing organism. The most essential point of difference is one of those seemingly slight ones, the importance of which we are too apt to overlook. It was like the drop of blood in the wrong place, which some one has told us makes all the difference between a philosopher and a maniac. It was all the difference between a living tree and a dead one, between an inert mass and a growing organism. The transition of knowledge from the dead to the living form must, in any complete review of the subject, be looked upon as the really great event of modern times.

9

Before this event the intellect was bound down by a scholasticism which regarded knowledge as a rounded whole, the parts of which were written in books and carried in the minds of learned men. The student was taught from the beginning of his work to look upon authority as the foundation of his beliefs. The older the authority the greater the weight it carried. So effective was this teaching that it seems never to have occurred to individual men that they had all the opportunities ever enjoyed by Aristotle of discovering truth, with the added advantage of all his knowledge to begin with. Advanced as was the development of formal logic, that practical logic was wanting which could see that the last of a series of authorities, every one of which rested on those which preceded it, could never form a surer foundation for any doctrine than that supplied by its original propounder.

The result of this view of knowledge was that, although during the fifteen centuries following the death of the geometer of Syracuse great universities were founded at which generations of professors expounded all the learning of their time, neither professor nor student ever suspected what latent possibilities of good were concealed in the most familiar operations of nature. Every one felt the wind blow, saw water boil, and heard the thunder crash, but never thought of investigating the forces here at play. Up to the middle of the fifteenth century the most acute observer could scarcely have seen the dawn of a new era.

In view of this state of things, it must be regarded as one of the most remarkable facts in evolutionary history that four or five men, whose mental constitution was either typical of the new order of things or who were powerful agents in bringing it about, were all born during the fifteenth century, four of them at least at so nearly the same time as to be contemporaries.

Leonardo da Vinci, whose artistic genius has charmed succeeding generations, was also the first practical engineer of his time, and the first man after Archimedes to make a substantial advance in developing the laws of motion. That the world was not prepared to make use of his scientific discoveries does not detract from the significance which must attach to the period of his birth.

Shortly after him was born the great navigator whose bold spirit was to make known a new world, thus giving to commercial enterprise that impetus which was so powerful an agent in bringing about a revolution in the thoughts of men.

The birth of Columbus was soon followed by that of Copernicus, the first after Aristarchus to demonstrate the true system of the world. In him more than in any of his contemporaries do we see the struggle between the old forms of thought and the new. It seems almost pathetic and is certainly most suggestive of the general view of knowledge taken at that time that, instead of claiming credit for bringing to light great truths before unknown, he made a labored attempt to show that, after all, there was nothing really new in his system, which he claimed to date from Pythagoras and Philolaus. In this connection it is curious that he makes no mention of Aristarchus, who I think will be regarded by conservative historians as his only demonstrated predecessor. To the hold of the older ideas upon his mind we must attribute the fact that in constructing his system he took great pains to make as little change as possible in ancient conceptions.

Luther, the greatest thought-stirrer of them all, practically of the same generation with Copernicus, Leonardo, and Columbus, does not come in as a scientific investigator, but as the great loosener of chains which had so fettered the intellect of men that they dared not think otherwise than as the authorities thought.

Almost coeval with the advent of these intellects was the invention of printing with movable type. Gutenberg was born during the first decade of the century, and his associates and others credited with the invention not many years afterward. If we accept the principle on which I am basing my argument, that we should assign the first place to the birth of those psychic agencies which started men on new lines of thought, then surely was the fifteenth the wonderful century.

Let us not forget that, in assigning the actors then born to their places, we are not narrating history, but studying a special phase of evolution. It matters not for us that no university invited Leonardo to its halls, and that his science was valued by his contemporaries only as an adjunct to the art of engineering. The great fact still is that he was the first of mankind to propound laws of motion. It is not for anything in Luther's doctrines that he finds a place in our scheme. No matter for us whether they were sound or not. What he did toward the evolution of the scientific investigator was to show by his example that a man might question the best-established and most venerable authority and still live — still preserve his intellectual integrity — still command a hearing from nations and their rulers. It matters not for us whether Columbus ever knew that he had discovered a new continent. His work was to teach that neither hydra, chimera, nor abyss — neither divine injunction nor infernal machination — was in the way of men visiting every part of the globe, and that the problem of conquering the world reduced itself to one of sails and rigging, hull and compass. The better part of Copernicus was to direct man to a viewpoint whence he should see that the heavens were of like matter with the earth. All this done, the acorn was planted from which the oak of our civilization should spring. The mad quest for gold which followed the discovery of Columbus, the questionings which absorbed the attention of the learned, the indignation excited by the seeming vagaries of a Paracelsus, the fear and trembling lest the strange doctrine of Copernicus should undermine the faith of centuries, were all helps to the germination of the seed — stimuli to thought which urged it on to explore the new fields opened up to its occupation. This given, all that has since followed came out in regular order of development, and need be here considered only in those phases having a special relation to the purpose of our present meeting.

So slow was the growth at first that the sixteenth century may scarcely have recognized the inauguration of a new era. Torricelli and Benedetti were of the third generation after Leonardo, and Galileo, the first to make a substantial advance upon his theory, was born more than a century after him. Only two or three men appeared in a generation who, working alone, could make real progress in discovery, and even these could do little in leavening the minds of their fellow men with the new ideas.

Up to the middle of the seventeenth century an agent which all experience since that time shows to be necessary to the most productive intellectual activity was wanting. This was the attraction of like minds, making suggestions to each other, criticising, comparing, and reasoning. This element was introduced by the organization of the Royal Society of London and the Academy of Sciences of Paris.

The members of these two bodies seem like ingenious youth suddenly thrown into a new world of interesting objects, the purposes and relations of which they had to discover. The novelty of the situation is strikingly shown in the questions which occupied the minds of the incipient investigators. One natural result of British maritime enterprise was that the aspirations of the Fellows of the Royal Society were not confined to any continent or hemisphere. Inquiries were sent all the way to Batavia to know "whether there be a hill in Sumatra which burneth continually, and a fountain which runneth pure balsam." The astronomical precision with which it seemed possible that physiological operations might go on was evinced by the inquiry whether the Indians can so prepare that stupefying herb Datura that "they make it lie several days, months, years, according as they will, in a man's body without doing him any harm, and at the end kill him without missing an hour's time." Of this continent one of the inquiries was whether there be a tree in Mexico that yields water, wine, vinegar, milk, honey, wax, thread, and needles.

Among the problems before the Paris Academy of Sciences those of physiology and biology took a prominent place. The distillation of compounds had long been practiced, and the fact that the more spirituous elements of certain substances were thus separated naturally led to the question whether the essential essences of life might not be discoverable in the same way. In order that all might participate in the experiments, they were conducted in open session of the Academy, thus guarding against the danger of any one member obtaining for his exclusive personal use a possible elixir of life. A wide range of the animal and vegetable kingdom, including cats, dogs, and birds of various species, were thus analyzed. The practice of dissection was introduced on a large scale. That of the cadaver of an elephant occupied several sessions, and was of such interest that the monarch himself was a spectator.

To the same epoch with the formation and first work of these two bodies belongs the invention of a mathematical method which in its importance to the advance of exact science may be classed with the invention of the alphabet in its relation to the progress of society at large. The use of algebraic symbols to represent quantities had its origin before the commencement of the new era, and gradually grew into a highly developed form during the first two centuries of that era. But this method could represent quantities only as fixed. It is true that the elasticity inherent in the use of such symbols permitted of their being applied to any and every quantity; yet, in any one application, the quantity was considered as fixed and definite. But most of the magnitudes of nature are in a state of continual variation; indeed, since all motion is variation, the latter is a universal characteristic of all phenomena. No serious advance could be made in the application of algebraic language to the expression of physical phenomena until it could be so extended as to express variation in quantities, as well as the quantities themselves. This extension, worked out independently by Newton and Leibnitz, may be classed as the most fruitful of conceptions in exact science. With it the way was opened for the unimpeded and continually accelerated progress of the last two centuries.

The feature of this period which has the closest relation to the purpose of our coming together is the seemingly unending subdivision of knowledge into specialties, many of which are becoming so minute and so isolated that they seem to have no interest for any but their few pursuers. Happily science itself has afforded a corrective for its own tendency in this direction. The careful thinker will see that in these seemingly diverging branches common elements and common principles are coming more and more to light. There is an increasing recognition of methods of research, and of deduction, which are common to large branches, or to the whole of science. We are more and more recognizing the principle that progress in knowledge implies its reduction to more exact forms, and the expression of its ideas in language more or less mathematical. The problem before the organizers of this Congress was, therefore, to bring the sciences together, and seek for the unity which we believe underlies their infinite diversity.

The assembling of such a body as now fills this hall was scarcely possible in any preceding generation, and is made possible now only through the agency of science itself. It differs from all preceding international meetings by the universality of its scope, which aims to include the whole of knowledge. It is also unique in that none but leaders have been sought out as members. It is unique in that so many lands have delegated their choicest intellects to carry on its work. They come from the country to which our republic is indebted for a third of its territory, including the ground on which we stand; from the land which has taught us that the most scholarly devotion to the languages and learning of the cloistered past is compatible with leadership in the practical application of modern science to the arts of life; from the island whose language and literature have found a new field and a vigorous growth in this region; from the last seat of the holy Roman Empire; from the country which, remembering a monarch who made an astronomical observation at the Greenwich Observatory, has enthroned science in one of the highest places in its government; from the peninsula so learned that we have invited one of its scholars to come and tell us of our own language; from the land which gave birth to Leonardo, Galileo, Torricelli, Columbus, Volta — what an array of immortal names! — from the little republic of glorious history which, breeding men rugged as its eternal snowpeaks, has yet been the seat of scientific investigation since the day of the Bernoullis; from the land whose heroic dwellers did not hesitate to use the ocean itself to protect it against invaders, and which now makes us marvel at the amount of erudition compressed within its little area; from the nation across the Pacific, which, by half a century of unequaled progress in the arts of life, has made an important contribution to evolutionary science through demonstrating the falsity of the theory that the most ancient races are doomed to be left in the rear of the advancing age — in a word, from every great centre of intellectual activity on the globe I see before me eminent representatives of that world-advance in knowledge which we have met to celebrate. May we not confidently hope that the discussions of such an assemblage will prove pregnant of a future for science which shall outshine even its brilliant past?

Gentlemen and scholars all! You do not visit our shores to find great collections in which centuries of humanity have given expression on canvas and in marble to their hopes, fears, and aspirations. Nor do you expect institutions and buildings hoary with age. But as you feel the vigor latent in the fresh air of these expansive prairies, which has collected the products of human genius by which we are here surrounded, and, I may add, brought us together; as you study the institutions which we have founded for the benefit, not only of our own people, but of humanity at large; as you meet the men who, in

14

the short space of one century, have transformed this valley from a savage wilderness into what it is to-day — then may you find compensation for the want of a past like yours by seeing with prophetic eye a future world-power of which this region shall be the seat. If such is to be the outcome of the institutions which we are now building up, then may your present visit be a blessing both to your posterity and ours by making that power one for good to all mankind. Your deliberations will help to demonstrate to us and to the world at large that the reign of law must supplant that of brute force in the relations of the nations, just as it has supplanted it in the relations of individuals. You will help to show that the war which science is now waging against the sources of diseases, pain, and misery offers an even nobler field for the exercise of heroic qualities than can that of battle. We hope that when, after your all too fleeting sojourn in our midst, you return to your own shores, you will long feel the influence of the new air you have breathed in an infusion of increased vigor in pursuing your varied labors. And if a new impetus is thus given to the great intellectual movement of the past century, resulting not only in promoting the unification of knowledge, but in widening its field through new combinations of effort on the part of its votaries, the projectors, organizers, and supporters of this Congress of Arts and Science will be justified of their labors.

Wilhelm Ostwald
1853–1932

SECTION D — METHODOLOGY OF SCIENCE

(*Hall 6, September 22, 3 p. m.*)

CHAIRMAN: PROFESSOR JAMES E. CREIGHTON, Cornell University.
SPEAKERS: PROFESSOR WILHELM OSTWALD, University of Leipzig.
PROFESSOR BENNO ERDMANN, University of Bonn.
SECRETARY: DR. R. B. PERRY, Harvard University.

ON THE THEORY OF SCIENCE

BY WILHELM OSTWALD

(*Translated from the German by Dr. R. M. Yerkes, Harvard University*)

[**Wilhelm Ostwald,** Professor of Physical Chemistry, University of Leipzig, since 1887. **b.** September 2, 1853, Riga, Russia. **Grad.** Candidate Chemistry, 1877; Master Chemistry, 1878; Doctor Chemistry, Dorpat. **Dr. Hon.** Halle and Cambridge; Privy Councilor; Assistant, Dorpat, 1875–81; Regular Professor, Riga, 1881–87. Member various learned and scientific societies. **Author of** *Manual of General Chemistry; Electro Chemistry; Foundation of Inorganic Chemistry; Lectures on Philosophy of Nature; Artist's Letters; Essays and Lectures;* and many other noted works and papers on Chemistry and Philosophy.]

ONE of the few points on which the philosophy of to-day is united is the knowledge that the only thing completely certain and undoubted for each one is the content of his own consciousness; and here the certainty is to be ascribed not to the content of consciousness in general, but only to the momentary content.

This momentary content we divide into two large groups, which we refer to the inner and outer world. If we call any kind of content of consciousness an experience, then we ascribe to the outer world such experiences as arise without the activity of our will and cannot be called forth by its activity alone. Such experiences never arise without the activity of certain parts of our body, which we call sense organs. In other words, the outer world is that which reaches our consciousness through the senses.

On the other hand, we ascribe to our inner world all experiences which arise without the immediate assistance of a sense organ. Here, first of all, belong all experiences which we call remembering and thinking. An exact and complete differentiation of the two territories is not intended here, for our purpose does not demand that this task be undertaken. For this purpose the general orientation in which every one recognizes familiar facts of his consciousness is sufficient.

Each experience has the characteristic of uniqueness. None of us doubts that the expression of the poet "Everything is only repeated in life" is really just the opposite of the truth, and that in fact nothing is repeated in life. But to express such a judgment we must be in position to compare different experiences with each other, and this possibility rests upon a fundamental phenomenon of our consciousness, memory. Memory alone enables us to put various experiences in relation to each other, so that the question as to their likeness or difference can be asked.

We find the simpler relations here in the inner experiences. A certain thought, such as twice two is four, I can bring up in my consciousness as often as I wish, and in addition to the content of the thought I experience the further consciousness that I have already had this thought before, that it is familiar to me.

A similar but somewhat more complex phenomenon appears in the experiences in which the outer world takes part. After I have eaten an apple, I can repeat the experience in two ways. First, as an inner experience, I can remember that I have eaten the apple and by an effort of my will I can re-create in myself, although with diminished strength and intensity, a part of the former experience — the part which belonged to my inner world. Another part, the sense impression which belonged to that experience, I cannot re-create by an effort of my will, but I must again eat an apple in order to have a similar experience of this sort. This is a complete repetition of the experience to which the external world also contributes. Such a repetition does not depend altogether on my own powers, for it is necessary that I have an apple, that is, that certain conditions which are independent of me and belong to the outer world be fulfilled.

Whether the outer world takes part in the repetition of an experience or not has no influence upon the possibility of the content of consciousness which we call memory. From this it follows that this content depends upon the inner experience alone, and that we remember an external event only by means of its inner constituents. The mere repetition of corresponding sense impressions is not sufficient for this, for we can see the same person repeatedly without recognizing him, if the inner accompanying phenomena were so insignificant, as a result of lack of interest, that their repetition does not produce the content of consciousness known as memory. If we see him quite frequently, the frequent repetition of the external impression finally causes the memory of the corresponding inner experience.

From this it results that for the "memory"-reaction a certain intensity of the inner experience is necessary. This threshold can be attained either at once or by continued repetition. The repetitions are the more effective the more rapidly they follow each other. From this we may conclude that the memory-value of an experience, or its capacity for calling forth the "memory"-reaction by repetition, decreases with the lapse of time. Further, we must consider the fact mentioned above, that an experience is never exactly repeated, and that therefore the "memory"-reaction occurs even where there is only resemblance or partial agreement in place of complete agreement. Here, too, there are different degrees; memory takes place more easily the more perfectly the two experiences agree, and *vice versa.*

If we look at these phenomena from the physiological side, we may say we have two kinds of apparatus or organs, one of which does not depend upon our will, whereas the other does. The former are the sense organs, the latter constitutes the organ of thought. Only the activities of the latter constitute our experiences or the content of our consciousness.

The activities of the former may call forth the corresponding processes of the latter, but this is not always necessary. Our sense organs can be influenced without our "noticing" it, that is, without the thinking apparatus being involved. An especially important reaction of the thinking apparatus is memory, that is, the consciousness that an experience which we have just had possesses more or less agreement with former experiences. With reference to the organ of thought, it is the expression of the general physiological fact that every process influences the organ in such a way that it has a different relation to the repetition of this process, from the first time, and moreover that the repetition is rendered easier. This influence decreases with time.

It is chiefly upon these phenomena that experience rests. Experience results from the fact that all events consist of a complete series of simultaneous and successive components. When a connection between some of those parts has become familiar to us by the repetition of similar occurrences (for instance, the succession of day and night), we do not feel such an occurrence as something completely new, but as something partially familiar, and the single parts or phases of it do not surprise us, but rather we anticipate their coming or expect them. From expectation to prediction is only a short step, and so experience enables us to prophesy the future from the past and present.

Now this is also the road to science; for science is nothing but systematized experience, that is, experience reduced to its simplest and clearest forms. Its purposes to predict from a part of a phenomenon which is known another part which is not yet known. Here it may be a question of spatial as well as of temporal phenomena. Thus the scientific zoölogist knows how to "determine," that is, to tell, from the skull of an animal, the nature of the other parts of the animal to which the skull belongs; likewise the astronomer is able to indicate the future situation of a planet from a few observations of its present situation; and the more exact the first observations were, the more distant the future for which he can predict. All such scientific predictions are limited, therefore, with reference to their number and their accuracy. If the skull shown to the zoölogist is that of a chicken, then he will probably be able to indicate the general characteristics of chickens, and also perhaps whether the chicken had a top-knot or not; but not its color, and only uncertainly its age and its size. Both facts, the possibility of prediction and its limitation in content and amount, are an expression for the two fundamental facts, that among our experiences there is similarity, but not complete agreement.

The foregoing considerations deserve to be discussed and extended in several directions. First, the objection will be made that a chicken or a planet is not an experience; we call them rather by the most general name of thing. But our knowledge of the chicken begins with the experiencing of certain visual impressions, to which are added, perhaps, certain impressions of hearing and touch. The sight impressions (to discuss these first) by no means completely agree. We see the chicken large or small, according to the distance; and according to its position and movement its outline is very different. As we have seen, however, these differences are continually grading into one other and do not reach beyond certain limits; we neglect to observe them and rest contented with the fact that certain other peculiarities (legs, wings, eyes, bill, comb, etc.) remain and do not change. The constant properties we group together as a thing, and the changing ones we call the states of this thing. Among the changing properties, we distinguish further those which depend upon us (for example, the distance) and those upon which we have no immediate influence (for instance, the position or motion): the first is called the subjective changeable part of our experience, while the second is called the objective mutability of the thing.

This omission of both the subjectively and objectively changeable portion of the experience in connection with the retention of the constant portion and the gathering together of the latter into a unity is one of the most important operations which we perform with our experiences. We call it the process of abstraction, and its product, the permanent unity, we call a concept. Plainly this procedure contains arbitrary as well as necessary factors. Arbitrary or accidental is the circumstance that quite different phases of a given experience come to consciousness according to our attention, the amount of practice we have had, indeed according to our whole intellectual nature. We may overlook constant factors and attend to changeable ones. The objective factors, however, become necessary as soon as we have noticed them; after we have seen that the chicken is black, it is not in our power to see it red. Accordingly, in general, our knowledge of that which agrees must be less than it actually could be, since we have not been able to observe every agreement, and our concept is always poorer in constituents at any given time than it might be. To seek out such elements of concepts as have been overlooked, and to prove that they are necessary factors of the corresponding experiences, is one of the never-ending tasks of science. The other case, namely, that elements have been received in the concept which do not prove to be constant, also happens, and leads to another task. One can then leave that element out of the concept, if further experiences show that the other elements are found in them, or one can form a new concept which contains the former elements, leaving out those that have been recognized as unessential. For a long time the white color belonged to the concept swan. When the Dutch black swans became known, it was possible either to drop the element white from the concept swan (as actually happened), or to make a new concept for the bird which is similar to the swan but black. Which choice is made in a given case is largely arbitrary, and is determined by considerations of expediency.

Into the formation of concepts, therefore, two factors are operative, an objective empirical factor, and a subjective or purposive factor. The fitness of a concept is seen in relation to its purpose, which we shall now consider.

The purpose of a concept is its use for prediction. The old logic set up the syllogism as the type of thought-activity, and its simplest example is the well-known

> All men are mortal,
> Caius is a man,
> Therefore Caius is mortal.

In general, the scheme runs

> To the concept M belongs the element B,
> C belongs under the concept M,
> Therefore the element B is found in C.

One can say that this method of reasoning is in regular use even to this day. It must be added, however, that this use is of a quite different nature from that of the ancients. Whereas formerly the setting up of the first proposition or the major premise was considered the most important thing, and the establishment of the second proposition or minor premise was thought to be a rather trifling matter, now the relation is reversed. The major premise contains the description of a concept, the minor makes the assertion that a certain thing belongs under this concept. What right exists for such an assertion? The most palpable reply would be, since all the elements of the concept M (including B) are found in C, C belongs under the concept M. Such a conclusion would indeed be binding, but at the same time quite worthless, for it only repeats the

21

minor premise.   Actually the method of reasoning is essentially different, for the minor premise is not obtained by showing that all the elements of the concept M are found in C, but only some of them. The conclusion is not necessary, but only probable, and the whole process of reasoning runs: Certain elements are frequently found together, therefore they are united in the concept M.   Certain of these elements are recognized in the thing C, therefore probably the other elements of the concept M will be found in C.

The old logic, also, was familiar with this kind of conclusion.   It was branded, however, as the worst of all, by the name of incomplete induction, since the absolute certainty demanded of the syllogism did not belong to its results. One must admit, however, that the whole of modern science makes use of no other form of reasoning than incomplete induction, for it alone admits of a prediction, that is, an indication of relations which have not been immediately observed.

How does science get along with the defective certainty of this process of reasoning?   The answer is, that the probability of the conclusion can run through all degrees from mere conjecture to the maximum probability, which is practically indistinguishable from certainty.   The probability is the greater the more frequently an incomplete induction of this kind has proven correct in later experience. Accordingly we have at our command a number of expressions which in their simplest and most general form have the appearance: If an element A is met within a thing, then the element B is also found in it (in spatial or temporal relationship).

If the relation is temporal, this general statement is known by some such name as the law of causality. If it is spatial, one talks of the idea (in the Platonic sense), or the type of the thing, of substance, etc.

From the considerations here presented we get an easy answer to many questions which are frequently discussed in very different senses.   First, the question concerning the general validity of the law of causality. All attempts to prove such a validity have failed, and there has remained only the indication that without this law we should feel an unbearable uncertainty in reference to the world. From this, however, we see very plainly that here it is merely a question of expediency. From the continuous flux of our experiences we hunt out those groups which can always be found again, in order to be able to conclude that if the element A is given, the element B will be present.   We do not find this relationship as "given," but we put it into our experiences, in that we consider the parts which correspond to the relationship as belonging together.

The very same thing may be said of spatial complexes. Such factors as are always, or at any rate often, found together are taken by us as "belonging together," and out of them a concept is formed which embraces these factors. A question as to the why has here, as with the temporal complexes, no definite meaning. There are countless things that happen together once to which we pay no attention because they happen only once or but seldom. The knowledge of the fact that such a single concurrence exists amounts to nothing, since from the presence of one factor it does not lead to a conclusion as to the presence of another, and therefore does not make possible prediction. Of all the possible, and even actual combinations, only those interest us which are repeated, and this arbitrary but expedient selection produces the impression that the world consists only of combinations that can be repeated; that, in other words, the law of causality or of the type is a general one. However general or limited application these laws have, is more a question of our skill in finding the constant combinations among those that are present than a question of objective natural fact.

Thus we see the development and pursuit of all sciences going on in such a way that on the one hand more and more constant combinations are discovered, and on the other hand more inclusive relations of this kind are found out, by means of which elements are united with each other which before no one had even tried to bring together. So sciences are increasing both in the sense of an increasing complication and in an increasing unification.

If we consider from this standpoint the development and procedure of the various sciences, we find a rational division of the sum total of science in the question as to the scope and multiplicity of the combinations or groups treated of in them. These two properties are in a certain sense antithetical. The simpler a complex is, that is, the fewer elements brought together in it, the more frequently it is met with, and *vice versa*. One can therefore arrange all the sciences in such a way that one begins with the least multiplicity and the greatest scope, and ends with the greatest multiplicity and the least scope. The first science will be the most general, and will therefore contain the most general and therefore the most barren concepts; the last will contain the most specific and therefore the richest.

What are these limiting concepts? The most general is the concept of *thing*, that is, any piece of experience, seized arbitrarily from the flux of our experiences, which can be repeated. The most specific and richest is the concept of *human intercourse*. Between the science of things and the science of human intercourse, all the other sciences are found arranged in regular gradation. If one follows out the scheme the following outline results:

1. Theory of order.
2. Theory of numbers, or arithmetic. ⎱
3. Theory of time. ⎰ Mathematics.
4. Theory of space, or geometry.

5. Mechanics. ⎫
6. Physics. ⎬ Energetics.
7. Chemistry. ⎭
8. Physiology. ⎫
9. Psychology. ⎬ Biology.
10. Sociology. ⎭

This table is arbitrary in so far as the grades assumed can be increased or diminished according to need. For example, mechanics and physics could be taken together; or between physics and chemistry, physical chemistry could be inserted. Likewise between physiology and psychology, anthropology might find a place; or the first five sciences might be united under mathematics. How one makes these divisions is entirely a practical question, which will be answered at any time in accordance with the purposes of division; and dispute concerning the matter is almost useless.

I should like, however, to call attention to the three great groups of mathematics, energetics, and biology (in the wider sense). They represent the decisive regulative thought which humanity has evolved, contributed up to this time, toward the scientific mastery of its experiences. Arrangement is the fundamental thought of mathematics. From mechanics to chemistry the concept of energy is the most important; and for the last three sciences it is the concept of life. Mathematics, energetics, and biology, therefore, embrace the totality of the sciences.

Before we enter upon the closer consideration of these sciences, it will be well to anticipate another objection which can be raised on the basis of the following fact. Besides the sciences named (and those which lie between them) there are many others, as geology, history, medicine, philology, which we find difficulty in arranging in the above scheme, which must, however, be taken into consideration in some way or other. They are often characterized by the fact that they stand in relation with several of the sciences named, but even more by the following circumstance. Their task is not, as is true of the pure sciences above named, the discovery of general relationships, but they relate rather to existing complex objects whose origin, scope, extent, etc., in short, whose temporal and spatial relationships they have to discover or to "explain." For this purpose they make use of relations which are placed at their disposal by the first-named pure sciences. These sciences, therefore, had better be called applied sciences. However, in this connection we should not think only or even chiefly of technical applications; rather the expression is used to indicate that the reciprocal relations of the parts of an object are to be called to mind by the application of the general rules found in pure science.

While in such a task the abstraction process of pure science is not applicable (for the omission of certain parts and the concentration upon others which is characteristic of these is excluded by the nature of the task), yet in a given case usually the necessity of bringing in various pure sciences for the purpose of explanation is evident.

Astronomy is one of these applied sciences. Primarily it rests upon mechanics, and in its instrumental portion, upon optics; in its present development on the spectroscopic side, however, it borrows considerably of chemistry. In like manner history is applied sociology and psychology. Medicine makes use of all the sciences before mentioned, up to psychology, etc.

It is important to get clearly in mind the nature of these sciences, since, on account of their compound nature, they resist arrangement amongst the pure sciences, while, on account of their practical significance, they still demand consideration. The latter fact gives them also a sort of arbitrary or accidental character, since their development is largely conditioned by the special needs of the time. Their number, speaking in general, is very large, since each pure science may be turned into an applied science in various ways; and since in addition we have combinations of two, three, or more sciences. Moreover, the method of procedure in the applied sciences is fundamentally different from that in the pure sciences. In the first it is a question of the greatest possible analysis of a single given complex into its scientifically comprehensible parts; while pure science, on the other hand, considers many complexes together in order to separate out from them their common element, but expressly disclaims the complete analysis of a single complex.

In scientific work, as it appears in practice, pure and applied science are by no means sharply separated. On the one hand the auxiliaries of investigations, such as apparatus, books, etc., demand of the pure investigator knowledge and application in applied science; and, on the other hand, the applied scientist is frequently unable to accomplish his task unless he himself becomes for the time being a pure investigator and ascertains or discovers the missing general relationships which he needs for his task. A separation and differentiation of the two forms of science was necessary, however, since the method and the aim of each present essential differences.

In order to consider the method of procedure of pure science more carefully, let us turn back to the table on pages 339, 340, and attend to the single sciences separately. The theory of arrangement was mentioned first, although this place is usually assigned to mathematics. However, mathematics has to do with the concepts of number and magnitude as fundamentals, while the theory of arrangement does not make use of these. Here the fundamental concept is rather the thing or object of which nothing more is demanded or considered than that it is a fragment of our experience which can be isolated and will remain so. It must not be an arbitrary combination; such a thing would have only momentary duration, and the task of science, to learn the unknown from the given, could not find application. Rather must this element have such a nature that it can be characterized and recognized again, that is, it must already have a conceptual nature. Therefore only parts of our experience which can be repeated (which alone can be objects of science) can be characterized as things or objects. But in saying this we have said all that was demanded of them. In other respects they may be just as different as is conceivable.

If the question is asked, What can be said scientifically about indefinite things of this sort? it is especially the relations of arrangement and association which yield an answer. If we call any definite combination of such things a group, we can arrange such a group in different ways, that is, we can determine for each thing the relation in which it is to stand to the neighboring thing. From every such arrangement result not only the relationships indicated, but a great number of new ones, and it appears that when the first relationships are given the others always follow in like manner. This, however, is the type of the scientific proposition or natural law (page 335). From the presence of certain relations of arrangement we can deduce the presence of others which we have not yet demonstrated.

To illustrate this fact by an example, let us think of the things arranged in a simple row, while we choose one thing as a first member and associate another with it as following it; with the latter another is associated, etc. Thereby the position of each thing in the row is determined only in relation to the immediately preceding thing. Nevertheless, the position of every member in the whole row, and therefore its relation to every other member, is determined by this. This is seen in a number of special laws. If we differentiate former and latter members we can formulate the proposition, among others, if B is a later member with reference to A, and C with reference to B, then C is also a later member with reference to A.

The correctness and validity of this proposition seems to us beyond all doubt. But this is only a result of the fact that we are able to demonstrate it very easily in countless single cases, and have so demonstrated it. We know only cases which correspond to the proposition, and have never experienced a contradictory case. To call such a proposition, however, a necessity of thinking, does not appear to me correct. For the expression necessity of thinking can only rest upon the fact that every time the proposition is thought, that is, every time one remembers its demonstration, its confirmation always arises. But every sort of false proposition is also thinkable. An undeniable proof of this is the fact that so much which is false is actually thought. But to base the proof for the correctness of a proposition upon the

26

impossibility of thinking its opposite is an impossible undertaking, because every sort of nonsense can be thought: where the proof was thought to have been given, there has always been a confusion of thought and intuition, proof or inspection.

With this one proposition of course the theory of order is not exhausted, for here it is not a question of the development of this theory, but of an example of the nature of the problems of science. Of the further questions we shall briefly discuss the problem of association.

If we have two groups A and B given, one can associate with every member of A one of B; that is, we determine that certain operations which can be carried on with the members of A are also to be carried on with those of B. Now we can begin by simply carrying out the association, member for member. Then we shall have one of three results: A will be exhausted while there are still members of B left, or B will be exhausted first, or finally A and B will be exhausted at the same time. In the first case we call A poorer than B; in the second B poorer than A; in the third both quantities are alike.

Here for the first time we come upon the scientific concept of equality, which calls for discussion. There can be no question of a complete identity of the two groups which have been denominated equal, for we have made the assumption that the members of both groups can be of any nature whatever. They can then be as different as possible, considered singly, but they are alike as groups. However I may arrange the members of A, I can make a similar arrangement of the members of B, since every member of A has one of B associated with it; and with reference to the property of arrangement there is no difference to be observed between A and B. If, however, A is poorer or richer than B, this possibility ceases, for then one of the groups has members to which none of the members in the other group corresponds; so that the operations carried out with these members cannot be carried out with those of the other group.

Equality in the scientific sense, therefore, means equivalence, or the possibility of substitution in quite definite operations or for quite definite relations. Beyond this the things which are called like may show any differences whatever. The general scientific process of abstraction is again easily seen in this special case.

On the basis of the definitions just given, we can establish further propositions. If group A equals B, and B equals C, then A also equals C. The proof of this is that we can relate every member of A to a corresponding member of B and by hypothesis no member will be left. Then C is arranged with reference to B, and here also no member is left. By this process every member of A, through the connecting link of a member of B, is associated with a member of C, and this association is preserved even if we cut out

27

the group B. Therefore A and C are equal. The same process of reasoning can be carried out for any number of groups.

Likewise it can be demonstrated that if A is poorer than B and B poorer than C, then A is also poorer than C. For in the association of B with A some members of B are left over by hypothesis, and likewise some members of C are left over if one associates C with B. Therefore in the association of C with A, not only those members are left over which could not be associated with B, but also those members of C which extend beyond B. This proposition can be extended to any number of groups, and permits the arrangement of a number of different groups in a simple series by beginning with the poorest and choosing each following so that it is richer than the preceding but poorer than the following. From the proposition just established, it follows that every group is so arranged with reference to all other groups that it is richer than all the preceding and poorer than all the following.[1]

In this derivation of scientific proposition or laws of the simplest kinds, the process of derivation and the nature of the result becomes particularly clear. We arrive at such a proposition by performing an operation and expressing the result of it. This expression enables us to avoid the repetition of the operation in the future, since in accordance with the law we can indicate the result immediately. Thus an abbreviation and therefore a facilitation of the problem is attained which is the more considerable the larger the number of operations saved.

If we have a number of equal groups, we know by the process of association that all of the operations with reference to arrangement which we can perform with one of them can be performed with all the others. It is sufficient, therefore, to determine the properties of arrangement of one of these groups in order to know forthwith the properties of all the others. This is an extremely important proposition, which is continually employed for the most various purposes. All speaking, writing, and reading rests upon the association of thoughts with sounds and symbols, and by arranging the signs in accordance with our thoughts we bring it to pass that our hearers or readers think like thoughts in like order. In a similar fashion we make use of various systems of formulæ in the different sciences, especially in the simpler sciences; and these formulæ we correlate with phenomena and use in place of the phenomena themselves, and can therefore derive from them certain characteristics of phenomena without being compelled to use the latter. The force of this process appears very strikingly in astronomy where, by the use of definite formulæ associated with the different heavenly bodies, we

---

[1] Equal groups cannot be distinguished here, and therefore represent only a group.

can foretell the future positions of these bodies with a high degree of approximation.

From the theory of order we come to the theory of number or arithmetic by the systematic arrangement or development of an operation just indicated (page 343). We can arrange any number of groups in such a way that a richer always follows a poorer. But the complex obtained in this manner is always accidental with reference to the number and the richness of its members. A regular and complete structure of all possible groups is evidently obtained only if we start from a group of one member or from a simple thing, and by the addition of one member at a time make further groups out of those that we have. Thus we obtain different groups arranged according to an increasing richness, and since we have advanced one member at a time, that is, made the smallest step which is possible, we are certain that we have left out no possible group which is poorer than the richest to which the operation has been carried.

This whole process is familiar; it gives the series of the positive whole numbers, that is, the cardinal numbers. It is to be noted that the concept of quantity has not yet been considered; what we have gained is the concept of number. The single things or members in this number are quite arbitrary, and especially they do not need to be alike in any manner. Every number forms a group-type, and arithmetic or the science of numbers has the task of investigating the properties of these different types with reference to their division and combination. If this is done in general form, without attention to the special amount of the number, the corresponding science is called algebra. On the other hand, by the application of formal rules of formation, the number system has had one extension after another beyond the territory of its original validity. Thus counting backward led to zero and to the negative numbers; the inversion of involution to the imaginary numbers. For the group-type of the positive whole numbers is the simplest but by no means the only possible one, and for the purpose of representing other manifolds than those which are met with in experience, these new types have proved themselves very useful.

At the same time the number series gives us an extremely useful type of arrangement. In the process of arising it is already ordered, and we make use of it for the purpose of arranging other groups. Thus, we are accustomed to furnish the pages in a book, the seats in a theatre, and countless other groups which we wish to make use of in any kind of order with the signs of the number series, and thereby we make the tacit assumption that the use of that corresponding group shall take place in the same order as the natural numbers follow each other. The ordinal numbers arising therefrom do not represent quantities, nor do they represent the only possible type

29

of arrangement, but they are again the simplest of all. We come to the concept of magnitude only in the theory of time and space. The theory of time has not been developed as a special science; on the contrary, what we have to say about time first appears in mechanics. Meantime we can present the fundamental concepts, which arise in this connection, with reference to such well-known characteristics of time that the lack of a special science of time is no disadvantage.

The first and most important characteristic of time (and of space, too) is that it is a continuous manifold; that is, every portion of time chosen can be divided at any place whatever. In the number series this is not the case; it can be divided only between the single numbers. The series one to ten has only nine places of division and no more. A minute, or a second, on the other hand, has an unlimited number of places of division. In other words, there is nothing in the lapse of any time which hinders us from separating or distinguishing in thought at any given instant the time which has elapsed till then from the following time. It is just the same with space, except that time is a simple manifold and space a threefold, continuous manifold.

Nevertheless, when we measure them, we are accustomed to indicate times and spaces with numbers. If we first examine, for example, the process of measuring a length, it consists in our applying to the distance to be measured a length conceived as unchangeable, the unit of measure, until we have passed over the distance. The number of these applications gives us the measure or magnitude of the distance. The result is that by the indication of arbitrarily chosen points upon the continuous distance, we place upon it an artificial discontinuity which enables us to associate it with the discontinuous number series.

A still further assumption, however, belongs to the concept of measuring, namely, that the parts of the distance cut off by the unit used as a measure be equal, and it is taken for granted that this requirement will be fulfilled to whatever place the unit of measure is shifted. As may be seen, this is a definition of equality carried further than the former, for one cannot actually replace a part of the distance by another in order to convince one's self that it has not changed. Just as little can one assert or prove that the unit of measure in changing its place in space remains of the same length; we can only say that such distances as are determined by the unit of measure in different places are declared or defined as equal. Actually, for our eye, the unit of measure becomes smaller in perspective the farther away from it we find ourselves.

From this example we see again the great contribution which arbitrariness or free choice has made to all our structure of science. We could develop a geometry in which distances which seem subjectively equal to our eye are called equal, and upon this assumption we would be able to develop a self-consistent system or science. Such a geometry, however, would have an extremely complex and impractical structure for objective purposes (as, for example, land measurement), and so we strive to develop a science as free as possible from subjective factors. Historically, we have before us a process of this sort in the astronomy of Ptolemy and that of Copernicus. The former corresponded to the subjective appearances in the assumption that all heavenly bodies revolved around the earth, but proved to be very complicated when confronted with the task of mastering these movements with figures. The latter gave up the subjective standpoint of the observer, who looked upon himself as the centre, and attained a tremendous simplification by placing the centre of revolution in the sun.

A few words are to be said here about the application of arithmetic and algebra to geometry. It is well known that under definite assumptions (coördinates), geometrical figures can be represented by means of algebraic formulæ, so that the geometrical properties of the figure can be deduced from the arithmetical properties of the formulæ, and *vice versa*. The question must be asked how such a close and univocal relationship is possible between things of such different nature. The answer is, that here is an especially clear case of association. The manifold of numbers is much greater than that of surface or space, for while the latter are determined by two or three independent measurements, one can have any number of independent number series working together. Therefore the manifold of numbers is arbitrarily limited to two or three independent series, and in so far determines their mutual relations (by means of the laws of cosine) that there results a manifold, corresponding to the spatial, which can be completely associated with the spatial manifold. Then we have two manifolds of the same manifold character, and all characteristics of arrangement and size of the one find their likeness in the other.

This again characterizes an extremely important scientific procedure which consists, namely, in constructing a formal manifold for the content of experience of a certain field, to which one attributes the same manifold character which the former possesses. Every science reaches by this means a sort of formal language of corresponding completeness, which depends upon how accurately the manifold character of the object is recognized and how judiciously the formulæ have been chosen. While in arithmetic and algebra this task has been performed fairly well (though by no means absolutely perfectly), the chemical formulæ, for instance, express only a relatively small part of the manifold to be represented; and in biology as far as sociology, scarcely the first attempts have been made in the accomplishment of this task.

Language especially serves as such a universal manifold to represent the manifolds of experience. As a result of its development from a time of less culture, it has by no means sufficient regularity and completeness to accomplish its purpose adequately and conveniently. Rather, it is just as unsystematic as the events in the lives of single peoples have been, and the necessity of expressing the endlessly different particulars of daily life has only allowed it to develop so that the correspondence between word and concept is kept rather indefinite and changeable, according to need within somewhat wide limits. Thus all work in those sciences which must make vital use of these means, as especially psychology and sociology, or philosophy in general, is made extremely difficult by the ceaseless struggle with the indefiniteness and ambiguity of language. An improvement of this condition can be effected only by introducing signs in place of words for the representation of concepts, as the progress of science allows it, and equipping these signs with the manifold which from experience belongs to the concept.

An intermediate position in this respect is taken by the sciences which were indicated above as parts of energetics. In this realm there is added to the concepts order, number, size, space, and time, a new concept, that of energy, which finds application to every single phenomenon in this whole field, just as do those more general concepts. This is due to the fact that a certain quantity, which is known to us most familiarly as mechanical work, on account of its qualitative transformability and quantitative constancy, can be shown to be a constituent of every physical phenomenon, that is, every phenomenon which belongs to the field of mechanics, physics, and chemistry. In other words, one can perfectly characterize every physical event by indicating what amounts and kinds of energy have been present in it and into what energies they have been transformed. Accordingly, it is logical to designate the so-called physical phenomena as energetical.

That such a conception is possible is now generally admitted. On the other hand, its expediency is frequently questioned, and there is at present so much the more reason for this because a thorough presentation of the physical sciences in the energetical sense has not yet been made. If one applies to this question the criterion of the scientific system given above, the completeness of the correspondence between the representing manifold and that to be represented, there is no doubt that all previous systematizations in the form of hypotheses which have been tried in these sciences are defective in this respect. Formerly, for the purpose of representing experiences, manifolds whose character corresponded to the character of the manifold to be represented only in certain salient points without consideration of any rigid agreement, indeed, even without definite question as to such an agreement, have been employed.

The energetical conception admits of that definiteness of representation which the condition of science demands and renders possible. For each special manifold character of the field a special kind of energy presents itself: science has long distinguished mechanical, electric, thermal, chemical, etc., energies. All of these different kinds hold together by the law of transformation with the maintenance of the quantitative amount, and in so far are united. On the other hand, it has been possible to fix upon the corresponding energetical expression for every empirically discovered manifold. As a future system of united energetics, we have then a table of possible manifolds of which energy is capable. In this we must keep in mind the fact that, in accordance with the law of the conservation, energy is a necessarily positive quantity which also is furnished with the property of unlimited possibility of addition; therefore, every particular kind of energy must have this character.

The very small manifold which seems to lack this condition is much widened by the fact that every kind of energy can be separated into two factors, which are only subject to the limitation that their product, the energy, fulfills the conditions mentioned while they themselves are much freer. For example, one factor of a kind of energy can become negative as well as positive; it is only necessary that at the same time the other factor should become negative, viz., positive.

Thus it seems possible to make a table of all possible forms of energy, by attributing all thinkable manifold characteristics to the factors of the energy and then combining them by pairs and cutting out those products which do not fulfill the above-mentioned conditions. For a number of years I have tried from time to time to carry out this programme, but I have not yet got far enough to justify publication of the results obtained.

If we turn to the biological sciences, in them the phenomenon of life appears to us as new. If we stick to the observed facts, keeping ourselves free from all hypotheses, we observe as the general characteristics of the phenomena of life the continuous stream of energy which courses through a relatively constant structure. Change of substance is only a part, although a very important part, of this stream. Especially in plants we can observe at first hand the great importance of energy in its most incorporeal form, the sun's rays. Along with this, self-preservation and development and reproduction, the begetting of offspring of like nature, are characteristic. All of these properties must be present in order that an organism may come into existence; they must also be present if the reflecting man is to be able by repeated experience to form a concept of any definite organism, whether of a lion or of a mushroom. Other organisms are met with which do not fulfill these conditions; on account of their rarity, however, they do not lead to a species concept, but are excluded from scientific consideration (except for special purposes) as deformities or monsters.

While organisms usually work with kinds of energy which we know well from the inorganic world, organs are found in the higher forms which without doubt cause or assist transfers of energy, but we cannot yet say definitely what particular kind of energy is active in them. These organs are called nerves, and their function is regularly that, after certain forms of energy have acted upon one end of them, they should act at the other end and release the energies stored up there which then act in their special manner. That energetical transformations also take place in the nerve during the process of nervous transmission can be looked upon as demonstrated. We shall thus be justified in speaking of a nerve energy, while leaving it undecided whether there is here an energy of a particular kind, or perhaps chemical energy, or finally a combination of several energies.

While these processes can be shown objectively by the stimulation of the nerve and its corresponding releasing reaction in the end apparatus (for instance, a muscle), we find in ourselves, connected with certain nervous processes, a phenomenon of a new sort which we call self-consciousness. From the agreement of our reactions with those of other people we conclude with scientific probability that they also have self-consciousness; and we are justified in making the same conclusion with regard to some higher animals. How far down something similar to this is present cannot be determined by the means at hand, since the analogy of organization and of behavior diminishes very quickly; but the line is probably not very long, in view of the great leap from man to animal. Moreover, there are many reasons for the view that the gray cortical substance in the brain, with its characteristic pyramidal cell, is the anatomical substratum of this kind of nervous activity.

The study of the processes of self-consciousness constitutes the chief task of psychology. To this science belong those fields which are generally allotted to philosophy, especially logic and epistemology, while æsthetics, and still more ethics, are to be reckoned with the social sciences.

The latter have to do with living beings in so far as they can be united in groups with common functions. Here in place of the individual mind appears a collective mind, which owing to the adjustment of the differences of the members of society shows simpler conditions than that. From this comes especially the task of the historical sciences. The happenings in the world accessible to us are conditioned partly by physical, partly by psychological factors, and both show a temporal mutability in one direction. Thus arises on the one hand a history of heaven and earth, on the other hand a history of organisms up to man.

All history has primarily the task of fixing past events through the effects which have remained from them. Where such are not accessible, only analogy is left, a very doubtful means for gaining a conception of those events. But it must be kept in mind that an event which has left no evident traces has no sort of interest for us, for our interest is directly proportional to the amount of change which that event has caused in what we have before us. The task of historical science is just as little exhausted, however, with the fixing of former events as, for instance, the task of physics with the establishment of a single fact, as the temperature of a given place at a given time. Rather the individual facts must serve to bring out the general characteristics of the collective mind, and the much-discussed historical laws are laws of collective psychology. Just as physical and chemical laws are deduced in order with their help to predict the course of future physical events (to be called forth either experimentally or technically), so should the historical laws contribute to the formation and control of social and political development. We see that the great statesmen of all time have eagerly studied history for this purpose, and from that we derive the assurance that there are historical laws in spite of the objections of numerous scholars.

After this brief survey, if we look back over the road we have come, we observe the following general facts. In every case the development of a science consists in the formation of concepts by certain abstractions from experience, and setting of these concepts in relation with each other so that a systematical control of certain sides of our experience is made possible. These relations, according to their generality and reliability, are called rules or laws. A law is the more important the more it definitely expresses concerning the greatest possible number of things, and the more accurately, therefore, it enables us to predict the future. Every law rests upon an incomplete induction, and is therefore subject to modification by experience. From this there results a double process in the development of science.

First, the actual conditions are investigated to find out whether, besides those already known, new rules or laws, that is, constant relations between individual peculiarities, cannot be discovered between them. This is the inductive process, and the induction is always an incomplete one on account of the limitlessness of all possible experience.

Immediately the relationship found inductively is applied to cases which have not yet been investigated. Especially such cases are investigated as result from a combination of several inductive laws. If these are perfectly certain, and the combination is also properly made, the result has claim to unconditional validity. This is the limit which all sciences are striving to reach. It has almost been reached in the simpler sciences: in mathematics and in certain parts of mechanics. This is called the deductive process.

In the actual working of every science the two methods of investigation are continually changing. The best means of finding new successful inductions is in the making of a deduction on a very insufficient basis, perhaps, and subsequently testing it in experience. Sometimes the elements of his deductions do not come into the investigator's consciousness; in such cases we speak of scientific instinct. On the other hand we have much evidence from great mathematicians that they were accustomed to find their general laws by the method of induction, by trying and considering single cases; and that the deductive derivation from other known laws is an independent operation which sometimes does not succeed until much later. Indeed there is to-day a number of mathematical propositions which have not yet reached the second stage and therefore have at present a purely inductive empirical character. The proportion of such laws in science increases very quickly with the rise in the scale (page 339).

Another peculiarity which may be mentioned here is that in the scale all previous sciences have the character of applied sciences (page 341) with reference to those which follow, since they are everywhere necessary in the technique of the latter, yet do not serve to increase their own field but are merely auxiliaries to the latter.

If we ask finally what influence upon the shaping of the future such investigations as those which have been sketched in outline above can have, the following can be said. Up till now it has been considered a completely uncontrollable event whether and where a great and influential man of science has developed. It is obvious that such a man is among the most costly treasures which a people (and, indeed, humanity) can possess. The conscious and regular breeding of such rarities has not been considered possible. While this is still the case for the very exceptional genius, we see in the countries of the older civilization, especially in Germany at present, a system of education in vogue in the universities by which a regular harvest of young scientific men is gained who not only have a mastery of knowledge handed down, but also of the technique of discovery. Thereby the growth of science is made certain and regular, and its pursuit is raised to a higher plane. These results were formerly attained chiefly by empirically and oftentimes by accidental processes. It is a task of scientific theory to make this activity also regular and systematic, so that success is no more dependent solely upon a special capacity for the founding of a "school" but can also be attained by less original minds. By the mastery of methods the way to considerably higher performances than he could otherwise attain will be open for the exceptionally gifted.

Benno Erdmann
1851–1921

THE CONTENT AND VALIDITY OF THE CAUSAL LAW

BY BENNO ERDMANN

(*Translated from the German by Professor Walter T. Marvin, Western Reserve University*)

[**Benno Erdmann,** Professor of Philosophy, University of Bonn, since 1898. b. October 5, 1851, Glogau in Schlesien, Germany. Ph.D.; Privy Councilor. Academical Lecturer, Berlin, 1876– ; Special Professor, Kiel, 1878–79; Regular Professor, *ibid.* 1879–84; *ibid.* Breslau, 1884–90; *ibid.* Halle, 1890–98. Member various scientific and learned societies. **Author of** *The Axioms of Geometry; Kant's Criticism; Logic; Psychological Researches on Reading* (together with Prof. Ramon Dodge); *The Psychology of the Child and the School; Historical Researches on Kant's Prolegomena,* and many other works and papers in Philosophy.]

WE have learned to regard the real, which we endeavor to apprehend scientifically in universally valid judgments, as a whole that is connected continuously in time and in space and by causation, and that is accordingly continuously self-evolving. This continuity of connection has the following result, namely, every attempt to classify the sum total of the sciences on the basis of the difference of their objects leads merely to representative types, that is, to species which glide into one another. We find no gaps by means of which we can separate sharply physics and chemistry, botany and zoölogy, political and economic history and the histories of art and religion, or, again, history, philology, and the study of the prehistoric.

As are the objects, so also are the methods of science. They are separable one from another only through a division into representative types; for the variety of these methods is dependent upon the variety of the objects of our knowledge, and is, at the same time, determined by the difference between the manifold forms of our thought, itself a part of the real, with its elements also gliding into one another.[1]

The threads which join the general methodology of scientific thought with neighboring fields of knowledge run in two main directions. In the one direction they make up a closely packed cable, whereas in the other their course diverges into all the dimensions of scientific thought. That is to say, first, methodology has its roots in logic, in the narrower sense, namely, in the science of the elementary forms of our thought which enter into the make-up of all scientific methods. Secondly, methodology has its source in the methods themselves which actually, and therefore technically, develop in the

[1] Cf. the author's "Theorie der Typeneinteilungen," *Philosophische Monatshefte,* vol. xxx, Berlin, 1894.

39

various fields of our knowledge out of the problems peculiar to those fields.

It is the office of scientific thought to interpret validly the objects that are presented to us in outer and inner perception, and that can be derived from both these sources. We accomplish this interpretation entirely through judgments and combinations of judgments of manifold sorts. The concepts, which the older logic regarded as the true elementary forms of our thinking, are only certain selected types of judgment, such stereotyped judgments as those which make up definitions and classifications, and which appear independent and fundamental because their subject-matter, that is, their intension or extension, is connected through the act of naming with certain words. Scientific methods, then, are the ways and means by which our thought can accomplish and set forth, in accordance with its ideal, this universally valid interpretation.

There belongs, accordingly, to methodology a list of problems which we can divide, to be sure only *in abstracto*, into three separate groups. First, methodology has to analyze the methods which have been technically developed in the different fields of knowledge into the elementary forms of our thinking from which they have been built up. Next to this work of *analyzing*, there comes a second task which may be called a *normative* one; for it follows that we must set forth and deduce systematically from their sources the nature of these manifold elements, their resulting connection, and their validity. To these two offices must be added a third that we may call *a potiori* a *synthetic* one; for finally we must reconstruct out of the elements of our thinking, as revealed by analysis, the methods belonging to the different fields of knowledge and also determine their different scope and validity.

The beginning of another conception of the office of methodology can be found in those thoughts which have become significant, especially in Leibnitz's fragments and drafts of a *calculus ratiocinator* or a *spécieuse générale*. The foregoing discussion has set aside all hope that these beginnings and their recent development may give, of the possibility of constructing the manifold possible methods *a priori*, that is, before or independent of experience. However, it remains entirely undecided, as it should in this our preliminary account of the office of general methodology, whether or not all methods of our scientific thought will prove to be ultimately but branches of one and the same universal method, a thought contained in the undertakings just referred to. Although modern empiricism, affiliated as it is with natural science, tends to answer this question in the affirmative even more definitely and dogmatically than any type of the older rationalism, still the question is one that can be decided only in the course of methodological research.

The conception of a methodology of scientific thought can be said to be almost as old as scientific thought itself; for it is already contained essentially, though undifferentiated, in the Socratic challenge of knowledge. None the less, the history of methodology, as the history of every other science, went through the course of which Kant has given a classical description. "No one attempts to construct a science unless he can base it on some idea; but in the elaboration of it the schema, nay, even the definition which he gives in the beginning of his science, corresponds very seldom to his idea, which, like a germ, lies hidden in the reason, and all the parts of which are still enveloped and hardly distinguishable even under microscopical observation." [1]

We are indebted to the Greek, and especially to the Platonic-Aristotelian philosophy for important contributions to the understanding of the deductive method of mathematical thought. It was precisely this trend of philosophic endeavor which, though furnishing for the most part the foundation of methodological doctrine well on into the seventeenth century, offered no means of differentiating the methods that are authoritative for our knowledge of facts. What Socrates was perhaps the first to call "induction," is essentially different, as regards its source and aim, from the inductive methods that direct our research in natural and mental science. For it is into these two fields that we have to divide the totality of the sciences of facts, the material sciences, let us call them, in opposition to the formal or mathematical sciences, — that is, if we are to do justice to the difference between sense and self-perception, or "outer" and "inner" perception.

Two closely connected forces especially led astray the methodological opinions regarding the material sciences till the end of the eighteenth century, and in part until the beginning of the nineteenth century. We refer, in the first place, to that direction of thought which gives us the right to characterize the Platonic-Aristotelian philosophy as a "concept philosophy;" namely, the circumstance that Aristotelian logic caused the "concept" to be set before the "judgment." In short, we refer to that tendency in thought which directs the attention not to the permanent in the world's occurrences, the uniform connections of events, but rather to the seemingly permanent in the things, their essential attributes or essences. Thus the concept philosophy, as a result of its tendency to hypostasize, finds in the abstract general concepts of things, the ideas, the eternal absolute reality that constitutes the foundation of things and is contained in them beside the accidental and changing properties. [2]

---

[1] Kant, *Kr. d. r. V.*, 2d ed., p. 862.
[2] According to Plato, it is true, the ideas are separated from the sensible things; they must be thought in a conceptual place, for the space of sense-perception is to

Here we have at once the second force which inspired the ancient methodology. These ideas, like the fundamentally real, constitute that which ultimately alone acts in all the coming into existence and the going out of existence of the manifold things. In the Aristotelian theory of causation, this thought is made a principle; and we formulate only what is contained in it, when we say that, according to it, the efficient and at the same time final causes can be deduced through mere analysis from the essential content of the effects; that, in fact, the possible effects of every cause can be deduced from the content of its definition. The conceptual determination of the causal relation, and with it in principle the sum total of the methods in the material sciences, becomes a logical, analytical, and deductive one. These sciences remain entirely independent of the particular content of experience as this broadens, and so do also the methods under discussion.

As a consequence, every essential difference between mathematical thought and the science of causes is done away with in favor of a rationalistic construction of the methods of material science. Accordingly, throughout the seventeenth century, the ideal of all scientific method becomes, not the inductive method that founded the new epoch of the science of to-day, but the deductive mathematical method applied to natural scientific research. The flourish of trumpets with which Francis Bacon hailed the onslaught of the inductive methods in the natural science of the time, helped in no way; for he failed to remodel the traditional, Aristotelian-Scholastic conception of cause, and, accordingly, failed to understand both the problem of induction and the meaning of the inductive methods of the day.[1] Descartes, Hobbes, Spinoza, and related thinkers develop their *mathesis universalis* after the pattern of geometrical thinking. Leibnitz tries to adapt his *spécieuse générale* to the thought of mathematical analysis. The old methodological conviction gains its clear-cut expression in Spinoza's doctrine: "*Aliquid efficitur ab aliqua re*" means "*aliquid sequitur ex ejus definitione.*"

The logically straight path is seldom the one taken in the course of the history of thought. The new formulation and solution of problems influence us first through their evident significance and consequences, not through the traditional presuppositions upon which they are founded. Thus, in the middle of the seventeenth century, when insight into the precise difference between mental and physical events gave rise to pressing need for its definite formulation, no question arose concerning the dogmatic presupposition

be understood as non-being, matter. The things revealed to sense, however, occupy a middle position between being and non-being, so that they partake of the ideas. In this sense, the statement made above holds also of the older view of the concept philosophy.

[1] Cf. the articles on Francis Bacon by Chr. Sigwart in the *Preussische Jahrbücher*, XII, 1863, and XIII, 1864.

of a purely logical (*analytisch*) relationship between cause and effect; but, on the contrary, this presupposition was then for the first time brought clearly before consciousness. It was necessary to take the roundabout way through occasionalism and the preëstablished harmony, including the latter's retreat to the omnipotence of God, before it was possible to raise the question of the validity of the presupposition that the connection between cause and effect is analytic and rational.

Among the leading thinkers of the period this problem was recognized as the cardinal problem of contemporaneous philosophy. It is further evidence how thoroughly established this problem must have been among the more deeply conceived problems of the time in the middle of the eighteenth century, that Hume and Kant were forced to face it, led on, seemingly independently of each other, and surely from quite different presuppositions and along entirely different ways. The historical evolution of that which from the beginning has seemed to philosophy the solving of her true problem has come to pass in a way not essentially different from that of the historical evolution in all other departments of human knowledge. Thus, in the last third of the seventeenth century, Newton and Leibnitz succeeded in setting forth the elements of the infinitesimal calculus; and, in the fifth decade of the nineteenth century, Robert Mayer, Helmholtz, and perhaps Joule, formulated the law of the conservation of energy. In one essential respect Hume and Kant are agreed in the solution of the new, and hence contemporaneously misunderstood, problem. Both realized that the connection between the various causes and effects is not a rational analytic, but an empirical synthetic one. However, the difference in their presuppositions as well as method caused this common result to make its appearance in very different light and surroundings. In Hume's empiricism the connection between cause and effect appears as the mere empirical result of association; whereas in Kant's rationalism this general relation between cause and effect becomes the fundamental condition of all possible experience, and is, as a consequence, independent of all experience. It rests, as a means of connecting our ideas, upon an inborn uniformity of our thought.

Thus the way was opened for a fundamental separation of the inductive material scientific from the deductive mathematical method. For Hume mathematics becomes the science of the relations of ideas, as opposed to the sciences of facts. For Kant philosophical knowledge is the knowledge of the reason arising from concepts, whereas the mathematical is that arising from the construction of concepts. The former, therefore, studies the particular only in the universal; the latter, the universal in the particular, nay, rather in the individual.

Both solutions of the new problem which in the eighteenth century supplant the old and seemingly self-evident presupposition, appear accordingly embedded in the opposition between the rationalistic and empiristic interpretation of the origin and validity of our knowledge, the same opposition that from antiquity runs through the historical development of philosophy in ever new digressions.

Even to-day the question regarding the meaning and the validity of the causal connection stands between these contrary directions of epistemological research; and the ways leading to its answer separate more sharply than ever before. It is therefore more pressing in our day than it was in earlier times to find a basis upon which we may build further epistemologically and therefore methodologically. The purpose of the present paper is to seek such a basis for the different methods employed in the sciences of facts.

As has already been said, the contents of our consciousness, which are given us immediately in outer and inner perception, constitute the raw material of the sciences of facts. From these various facts of perception we derive the judgments through which we predict, guide, and shape our future perception in the course of possible experience. These judgments exist in the form of reproductive ideational processes, which, if logically explicit, become *inductive inferences* in the broader sense. These inferences may be said to be of two sorts, though fundamentally only two sides of one and the same process of thought; they are in part analogical inferences and in part *inductive inferences in the narrower sense.* The former infers from the particular in a present perception, *which in previous perceptions was uniformly connected with other particular contents of perception,* to a particular that resembles *those other contents of perception.* In short, they are inferences from a particular to a particular. After the manner of such inferences we logically formulate, for example, the reproductive processes, whose conclusions run: "This man whom I see before me, is attentive, feels pain, will die;" "this meteor will prove to have a chemical composition similar to known meteors, and also to have corresponding changes on its surface as the result of its rapid passage through our atmosphere." The inductive inferences in the narrower sense argue, on the contrary, from the perceptions of a series of uniform phenomena to a universal, which includes the given and likewise all possible cases, in which a member of the particular content of the earlier perceptions is presupposed as given. In short, they are conclusions from a particular to a universal that is more extensive than the sum of the given particulars. For example: "All men have minds, will die;" "all meteoric stones will prove to have this chemical composition and those changes of surface."

There is no controversy regarding the inner similarity of both these types of inference or regarding their outward structure; or, again, regarding their outward difference from the deductive inferences, which proceed not from a particular to a particular or general, but from a general to a particular.

There is, however, difference of opinion regarding their inner structure and their inner relation to the deductive inferences. Both questions depend upon the decision regarding the meaning and validity of the causal relation. The contending parties are recruited essentially from the positions of traditional empiricism and rationalism and from their modern offshoots.

We maintain first of all:

1. The *presupposition* of all inductive inferences, from now on to be taken in their more general sense, is, that the contents of perception are given to us *uniformly* in repeated perceptions, that is, in uniform components and uniform relations.

2. The *condition* of the validity of the inductive inferences lies in the thoughts that *the same causes will be present* in the unobserved realities as in the observed ones, and that *these same causes will bring forth the same effects*.

3. The *conclusions* of all inductive inferences have, logically speaking, purely *problematic* validity, that is, their contradictory opposite remains equally thinkable. They are, accurately expressed, merely *hypotheses*, whose validity needs verification through future experience.

The first-mentioned *presupposition* of inductive inference must not be misunderstood. The paradox that nothing really repeats itself, that each stage in nature's process comes but once, is just as much and just as little justified as the assertion, everything has already existed. It does not deny the fact that we can discriminate in the contents of our perceptions the uniformities of their components and relations, in short, that similar elements are present in these ever new complexes. This fact makes it possible that our manifold perceptions combine to make up one continuous experience. Even our paradox presupposes that the different contents of our perceptions are comparable with one another, and reveal accordingly some sort of common nature. All this is not only a matter of course for empiricism, which founds the whole constitution of our knowledge upon habits, but must also be granted by every rationalistic interpretation of the structure of knowledge. Every one that is well informed knows that what we ordinarily refer to as facts already includes a theory regarding them. Kant judges in this matter precisely as Hume did before him and Stuart Mill after him. "If cinnabar were sometimes red and sometimes black, sometimes light and sometimes heavy, if a man could be changed now into this, now into another animal shape, if on the longest day the fields were sometimes covered with fruit, sometimes with ice and snow, the faculty of my empirical imagination would never be in a position, when representing red color, to think of heavy cinnabar." [1]

The assumption that in recurring perceptions similar elements of content, as well as of relation, are given, is a necessary condition of the possibility of experience itself, and accordingly of all those processes of thought which lead us, under the guidance of previous perceptions, from the contents of one given perception to the contents of possible perceptions.

A tradition from Hume down has accustomed us to associate the relation of cause and effect not so much with the uniformity of coexistence as with the uniformity of sequence. Let us for the present keep to this tradition. Its first corollary is that the relation of cause and effect is to be sought in the uninterrupted flow and connection of events and changes. The cause becomes the uniformly preceding event, the constant *antecedens*, the effect the uniformly following, the constant *consequens*, in the course of the changes that are presented to consciousness as a result of foregoing changes in our sensorium.

According to this tradition that we have taken as our point of departure, the uniformity of the sequence of events is a necessary presupposition of the relation between cause and effect. This uniformity is given us as an element of our experience; for we actually find uniform successions in the course of the changing contents of perception. Further, as all our perceptions are in the first instance sense-perceptions, we may call them the sensory presupposition of the possibility of the causal relation.

In this presupposition, however, there is much more involved than the name just chosen would indicate. The uniformity of sequence lies, as we saw, not in the contents of perception as such, which are immediately given to us. It arises rather through the fact that, in the course of repeated perceptions, we apprehend through abstraction the uniformities of their temporal relation. Moreover, there lie in the repeated perceptions not only uniformities of sequence, but also uniformities of the qualitative content of the successive events themselves, and these uniformities also must be apprehended through abstraction. Thus these uniform contents of perception make up series of the following form:

$$a_1 \rightarrow b_1$$
$$a_2 \rightarrow b_2$$
$$" \quad "$$
$$" \quad "$$
$$" \quad "$$
$$a_n \rightarrow b_n$$

[1] Kant, *Kr. d. r. V.*, 1st ed., pp. 100 f.

46

The presupposition of the possibility of the causal relations includes, therefore, more than mere perceptive elements. It involves the relation of different, if you will, of peculiar contents of perception, by virtue of which we recognize $a_2 \to b_2$ . . . $a_n \to b_n$ as events that resemble one another and the event $a_1 \to b_1$ qualitatively as well as in their sequence. There are accordingly involved in our presupposition *reproductive* elements which indicate the action of memory. In order that I may in the act of perceiving $a_3 \to b_3$ apprehend the uniformity of this present content with that of $a_2 \to b_2$ and $a_1 \to b_1$, these earlier perceptions must in some way, perhaps through memory,[1] be revived with the present perception.

In this reproduction there is still a further element, which can be separated, to be sure only *in abstracto*, from the one just pointed out. The present revived content, even if it is given in memory as an independent mental state, is essentially different from the original perception. It differs in all the modifications in which the memory of lightning and thunder could differ from the perception of their successive occurrence, or, again, the memory of a pain and the resulting disturbance of attention could differ from the corresponding original experience. However, as memory, the revived experience presents itself as a picture of that which has been previously perceived. Especially is this the case in memory properly so called, where the peculiar space and time relations individualize the revived experience. If we give to this identifying element in the associative process a logical expression, we shall have to say that there is involved in revival, and especially in memory, an awareness that the present ideas recall the same content that was previously given us in perception. To be sure, the revival of the content of previous perceptions does not have to produce ideas, let alone memories. Rapid, transitory, or habitual revivals, stimulated by associative processes, can remain unconscious, that is, they need not appear as ideas or states of consciousness. Stimulation takes place, but consciousness does not arise, provided we mean by the term "consciousness" the genus of our thoughts, feelings, and volitions. None the less it must not be forgotten that this awareness of the essential identity of the present revived content with that of the previous perception can be brought about in every such case of reproduction. How all this takes place is not our present problem.

We can apply to this second element in the reproductive process, which we have found to be essential to the causal relation, a Kantian

---

[1] It is not our present concern to ascertain how this actually happens. The psychological presuppositions of the present paper are contained in the theory of reproduction that I have worked out in connection with the psychology of speech in the articles on "Die psychologischen Grundlagen der Beziehungen zwischen Sprechen und Denken," *Archiv für systematische Philosophie*, II, III, und VII; cf. note 1, page 151.

term, "Recognition." This term, however, is to be taken only in the sense called for by the foregoing statements; for the rationalistic presuppositions and consequences which mark Kant's "Synthesis of Recognition" are far removed from the present line of thought.

We may, then, sum up our results as follows: In the presupposition of a uniform sequence of events, which we have accepted from tradition as the necessary condition of the possibility of the causal relation, there lies the thought that the contents of perception given us through repeated sense stimulation are related to one another through a reproductive recognition.

The assumption of such reproductive recognition is not justified merely in the cases so far considered. It is already necessary in the course of the individual perceptions a and b, and hence in the apprehension of an occurrence. It makes the sequence itself in which a and b are joined possible; for in order to apprehend b as following upon a, in case the perception of a has not persisted in its original form, a must be as far revived and recognized upon b's entrance into the field of perception as it has itself passed out of that field. Otherwise, instead of b following upon a and being related to a, there would be only the relationless change from a to b. This holds generally and not merely in the cases where the perception of a has disappeared before that of b begins, for example, in the case of lightning and thunder, or where it has in part disappeared, for example, in the throwing of a stone.

We have represented a as an event or change, in order that uniform sequences of events may alone come into consideration as the presupposition of the causal relation. But every event has its course in time, and is accordingly divisible into many, ultimately into infinitely many, shorter events. Now if b comes only an infinitely short interval later than a, and by hypothesis it must come later than a, then a corresponding part of a must have disappeared by the time b appears. But the infinitesimal part of a perception is just as much out of all consideration as would be an infinitely long perception; all which only goes to show that we have to substitute intervals of finite length in place of this purely conceptual analysis of a continuous time interval. This leaves the foregoing discussion as it stands. If b follows a after a perceptible finite interval, then the flow or development of a by the time of b's appearance must have covered a course corresponding to that interval; and all this is true even though the earlier stages of a remain unchanged throughout the interval preceding b's appearance. The present instant of flow is distinct from the one that has passed, even though it takes place in precisely the same way. The former, not the latter, gives the basis of relation which is here required, and therefore the former must be reproduced and recognized. This thought also is included in the foregoing summary of what critical analysis shows to be involved in the presupposition of a uniform sequence.

In all this we have already abandoned the field of mere perception which gave us the point of departure for our analysis of uniform sequence. We may call the changing course of perception only in the narrower meaning the sensory presupposition of the causal relation. In order that these changing contents of perception may be known as like one another, as following one another, and as following one another uniformly, they must be related to one another through a recognitive reproduction.

Our critical analysis of uniform sequence is, however, not yet complete. To relate to one another the contents of two ideas always requires a process at once of identifying and of differentiating, which makes these contents members of the relation, and which accordingly presupposes that our attention has been directed to each of the two members as well as to the relation itself — in the present case, to the sequence. Here we come to another essential point. We should apply the name "thought" to every ideational process in which attention is directed to the elements of the mental content and which leads us to identify with one another, or to differentiate from one another, the members of this content.[1] The act of relating, which knows two events as similar, as following one another, indeed, as following one another uniformly, is therefore so far from being a sensation that it must be claimed to be an act of thinking. The uniformity of sequence of a and b is therefore an act of relating on the part of our thought, so far as this becomes possible solely through the fact that we at one and the same time identify with one another and differentiate from one another a as cause and b as effect. We say "at one and the same time," because the terms identifying and differentiating are correlatives which denote two different and opposing sides of one and the same ideational process viewed logically. Accordingly, there is here no need of emphasizing that the act of relating, which enables us to think a as cause and b as effect, is an act of thought also, because it presupposes on our part an act of naming which raises it to being a component of our formulated and discursive thought. We therefore *think* a as cause and b as effect in that we apprehend the former as uniform *antecedens* and the latter as uniform *consequens*.

Have we not the right, after the foregoing analysis, to interpret the uniform sequence of events solely as the *necessary* presupposition of the causal relation? Is it not at the same time the *adequate* presupposition? Yes, is it not the causal relation itself? As we know, empiricism since Hume has answered the last question in the

---

[1] Cf. the author's "Umrisse zur Psychologie des Denkens," in *Philosophische Abhandlungen Chr. Sigwart . . . gewidmet*, Tübingen, 1900.

affirmative, and rationalism since Kant has answered it in the negative.

We, too, have seemingly followed in our discussion the course of empiricism. At least, I find nothing in that discussion which a consistent empiricist might not be willing to concede; that is, if he is ready to set aside the psychological investigation of the actual processes which we here presuppose and make room for a critical analysis of the content of the relation of cause and effect.[1] However, the

---

[1] The difference between the two points of view can be made clearer by an illustration. The case that we shall analyze is the dread of coming into contact with fire. The psychological analysis of this case has to make clear the mental content of the dread and its causes. Such dread becomes possible only when we are aware of the burning that results from contact with fire. We could have learned to be aware of this either immediately through our own experience, or mediately through the communication of others' experience. In both cases it is a matter of one or repeated experiences. In all cases the effects of earlier experiences equal association and recall, which, in turn, result in recognition. The recognition explaining the case under discussion arises thus. The present stimuli of visual perception arouse the retained impressions of previous visual perceptions of fire and give rise to the present perception (apperception) by fusing with them. By a process of interweaving, associations are joined to this perception. The apperceptively revived elements which lie at the basis of the content of the perception are interwoven by association with memory elements that retain the additional contents of previous perceptions of fire, viz., the burning, or, again, are interwoven with the memory elements of the communications regarding such burning. By means of this interweaving, the stimulation of the apperceptive element transmits itself to the remaining elements of the association complex. The character of the association is different under different conditions. If it be founded only upon one experience, then there can arise a memory or a recall, in the wider sense, of the foregoing content of the perception and feeling at the time of the burning, or, again, there can arise a revival wherein the stimulated elements of retention remain unconscious. Again, the words of the mother tongue that denote the previous mental content, and which likewise belong to the association complex (the apperceiving mass, in the wider sense), can be excited in one of these three forms and in addition as abstract verbal ideas. Each one of these forms of verbal discharge can lead to the innervations of the muscles involved in speech, which bring about some sort of oral expression of judgment. Each of these verbal reproductions can be connected with each of the foregoing sensory (*sachlichen*) revivals. Secondly, if the association be founded upon repeated perceptions on the part of the person himself, then all the afore-mentioned possibilities of reproduction become more complicated, and, in addition, the mental revivals contain, more or less, only the common elements of the previous perceptions, i. e., reappear in the form of abstract ideas or their corresponding unconscious modifications. In the third case the association is founded upon a communication of others' experience. For the sake of simplicity, let this case be confined to the following instance. The communication consisted in the assertion: "All fire will burn upon contact." Moreover, this judgment was expressed upon occasion of imminent danger of burning. There can then arise, as is perhaps evident, all the possibilities mentioned in the second case, only that here there will be a stronger tendency toward verbal reproduction and the sensory reproduction will be less fixed.

In the first two cases there was connected with the perception of the burning an intense feeling of pain. In the third the idea of such pain added itself to the visual perception of the moment. The associated elements of the earlier mental contents belong likewise to the apperceiving mass excited at the moment, in fact to that part of it excited by means of association processes, or, as we can again say, depending upon the point from which we take our view, the associative or apperceptive completion of the content of present perception. If these pain elements are revived as memories, i. e., as elements in consciousness, they give rise to a new disagreeable feeling, which is referred to the possible coming sensation of burning. If the mental modifications corresponding to these pain elements remain unconscious, as is often possible, there arises none the less the same result as regards our feeling, only with less intensity. This feeling tone we call the dread.

decision of the question, whether or not empiricism can determine exhaustively the content that we think in the causal relation, depends upon other considerations than those which we have until now been called upon to undertake. We have so far only made clear what every critical analysis of the causal relation has to concede to empiricism. In reality the empiristic hypothesis is inadequate. To be sure,

As a result of the sum total of the revivals actual and possible, there is finally produced, according to the particular circumstances, either a motor reaction or an inhibitant of such reaction. Both innervations can take place involuntarily or voluntarily.

The critical analysis of the fact that we dread contact with fire, even has another purpose and accordingly proceeds on other lines. It must make clear under what presuppositions the foresight that lies at the basis of such dread is valid for future experience. It must then formulate the actual process of revival that constitutes the foundation of this feeling as a series of judgments, from which the meaning and interconnection of the several judgments will become clear. Thus the critical analysis must give a logical presentation of the apperceptive and associative processes of revival.

For this purpose the three cases of the psychological analysis reduce themselves to two: viz., first, to the case in which an immediate experience forms the basis, and secondly, to that in which a variety of similar mediately or immediately communicated experiences form such basis.

In the first of these logically differentiated cases, the transformation into the speech of formulated thought leads to the following inference from analogy:

Fire A burned.
Fire B is similar to fire A.

Fire B will burn.

In the second case there arises a syllogism of some such form as:

All fire causes burning upon contact.
This present phenomenon is fire.

This present phenomenon will cause burning upon contact.

Both premises of this syllogism are inductive inferences, whose implicit meaning becomes clear when we formulate as follows:

All heretofore investigated instances of fire have burned, therefore all fire burns.
The present phenomenon manifests some properties of fire, will consequently have all the properties thereof.

The present phenomenon will, in case of contact, cause burning.

The first syllogism goes from the particular to the particular. The second proves itself to be (contrary to the analysis of Stuart Mill) an inference that leads from the general to the particular. For the conclusion is the particular of the second parts of the major and minor premises; and these second parts of the premises are inferred from their first parts in the two possible ways of inductive inference. The latter do not contain the case referred to in the conclusion, but set forth the conditions of carrying a result of previous experience over to a new case with inductive probability, in other words, the conditions of making past experience a means of foreseeing future experience. It would be superfluous to give here the symbols of the two forms of inductive inference.

We remain within the bounds of logical analysis, if we state under what conditions conclusions follow necessarily from their premises, viz., the conclusions of arguments from analogy and of syllogisms in the narrower sense, as well as those of the foregoing inductive arguments. For the inference from analogy and the two forms of inductive inference, these conditions are the presuppositions already set forth in the text of the present paper, that in the as yet unobserved portion of reality the like causes will be found and they will give rise to like effects. For the syllogism they are the thought that the predicate of a predicate is the (mediate) predicate of the subject. Only the further analysis of these presuppositions, which is undertaken in the text, leads to critical considerations in the narrower sense.

the proof of this inadequacy is not to be taken from the obvious argument which Reid raised against the empiricism of Hume, and which compelled Stuart Mill in his criticism of that attack [1] to abandon his empiristic position at this point. No doubt the conclusion to which we also have come for the time being, goes much too far, the conclusion that the cause is nothing but the uniform *antecedens* and the effect merely the uniform *consequens*. Were it true, as we have hitherto assumed, that every uniformly preceding event is to be regarded as cause and every uniformly following event as effect, then day must be looked upon as cause of night and night as cause of day.

Empiricism can, however, meet this objection without giving up its position; in fact, it can employ the objection as an argument in its favor; for this objection affects only the manifestly imperfect formulation of the doctrine, not the essential arguments.

It should have been pointed out again and again in the foregoing exposition that only in the first indiscriminating view of things may we regard the events given us in perception as the basis of our concepts of cause and effect. All these events are intricately mixed, those that are given in self perception as well as those given in sense perception. The events of both groups flow along continuously. Consequently, as regards time, they permit a division into parts, which division proceeds, not indeed for our perception, but for our scientific thought. in short, conceptually, into infinity. The events of sense perception permit also conceptually of infinite division in their spatial relations.

It is sufficient for our present purpose, if we turn our attention to the question of divisibility in time. This fact of divisibility shows that the events of our perception, which alone we have until now brought under consideration, must be regarded as systems of events. We are therefore called upon to apportion the causal relations among the members of these systems. Only for the indiscriminating view of our practical *Weltanschauung* is the perceived event a the cause of the perceived event b. The more exact analysis of our theoretical apprehension of the world compels us to dissect the events a and b into the parts a_α, a_β, a_γ, $-b_\alpha$, b_β, b_γ, and, where occasion calls for it, to continue the same process in turn for these and further components. We have accordingly to relate those parts to one another as causes and effects which, from the present standpoint of analysis, follow one another uniformly and *immediately*, viz., follow one another so that from this standpoint no other intervening event must be presupposed. In this way we come to have a *well-ordered experience*. The dispositions to such experience which reveal themselves within the field of practical thought taught man long before the beginning of scientific methods not to connect causally day and night with one another, but the rising and setting of the sun with day and night. The theoretical

---

[1] *A System of Logic, Ratiocinative and Inductive*, bk. III, ch. V, § 6.

analysis, indeed, goes farther. It teaches that in what is here summed up as rising of the sun and yonder as day, there lie again intricate elements requiring special attention, in our own day extending perhaps to the lines of thought contained in the electro-dynamic theory of light and of electrons. Still the ways of thought remain the same on all the levels of penetrating analysis. We have throughout to relate to one another as cause and effect those events which, in a well-ordered experience, must be regarded as following one another immediately. The cause is then the *immediate* uniform *antecedens*, the effect the *immediate* uniform *consequens*. Otherwise stated, the perceived events that we are accustomed, from the standpoint of the practical *Weltanschauung*, to regard as causes and effects, *e. g.*, lightning and thunder, from the theoretical apprehension of the world prove to be infinitely involved collections of events, whose elements must be related to one another as causes and effects in as far as they can be regarded as following one another immediately. No exception is formed by expressions of our rough way of viewing and describing which lead us without hesitation to regard as cause one out of the very many causes of an event, and this, too, not necessarily the immediate uniformly preceding event. All this lies rather in the nature of such a hasty view.

The present limitation of uniform sequence to cases of immediate sequence sets aside, then, the objection from which we started, in that it adopts as its own the essential point in question.

Moreover, the way that leads us to this necessary limitation goes farther: it leads to a strengthening of the empiristic position. It brings us to a point where we see that the most advanced analysis of intricate systems of events immediately given to us in perception as real nowhere reveals more than the simple fact of uniform sequence Again where we come to regard the intervals between the events that follow one another immediately as very short, there the uniformity of the time relation makes, it would seem, the events for us merely causes and effects; and as often as we have occasion to proceed to the smaller time differences of a higher order. the same process repeats itself; for we dissect the events that make up our point of departure into ever more complex systems of component events, and the coarser relations of uniform sequence into ever finer immediate ones. Nowhere, seemingly, do we get beyond the field of events in uniform sequence, which finally have their foundation in the facts of perception from which they are drawn. Thus there follows from this conceptual refinement of the point of departure only the truth that nothing connects the events as causes and effects except the immediate uniformity of sequence.

None the less, we have to think the empiristic doctrine to the bottom, if we desire to determine whether or not the hypothesis which it offers is really sufficient to enable us to deduce the causal relation. For this purpose let us remind ourselves that the question at issue is, whether or not this relation is merely a temporal connection of events that are given to us in perception or that can be derived from the data of perception.

Besides, let us grant that this relation is as thoroughly valid for the content of our experience as empiricism has always, and rationalism nearly always, maintained. We presuppose, therefore, as granted, that every event is to be regarded as cause, and hence, in the opposite time relation, as effect, mental events that are given to us in self perception no less than the physical whose source is our sense perception. In other words, we assume that the totality of events in our possible experience presents a closed system of causal series, that is, that every member within each of the contemporary series is connected with the subsequent ones, as well as with the subsequent members of all the other series, backward and forward as cause and effect; and therefore, finally, that every member of every series stands in causal relationship with every member of every other series. We do not then, for the present purpose, burden ourselves with the hypothesis which was touched upon above, that this connection is to be thought of as a continuous one, namely, that other members can be inserted *ad infinitum* between any two members of the series.

We maintain at the same time that there is no justification for separating from one another the concepts, causality and interaction. This separation is only to be justified through the metaphysical hypothesis that reality consists in a multitude of independently existing substances inherently subject to change, and that their mutual interconnection is conditioned by a common dependence upon a first infinite cause.[1] Every connection between cause and effect is mutual, if we assume with Newton that to every action there is an equal opposing reaction.

In that we bring the totality of knowable reality, as far as it is analyzable into events, under the causal relation, we may regard the statement that every event requires us to seek among uniformly preceding events for the sufficient causes of its own reality, namely, *the general causal law*, as the principle of all material sciences. For all individual instances of conformity to law which we can discover in the course of experience are from this point of view only special cases of the general universal conformity to law which we have just formulated.

[1] This doctrine began in the theological evolution of the Christian concept of God. It was first fundamentally formulated by Leibnitz. It is retained in Kant's doctrine of the *harmonia generaliter stabilita* and the latter's consequences for the critical doctrine of the *mundus intelligibilis*. Hence it permeates the metaphysical doctrines of the systems of the nineteenth century in various ways.

For the empiristic interpretation, the (general) causal law is only the highest genus of the individual cases of empirically synthetic relations of uniform sequence. Starting from these presuppositions, it cannot be other than a generalization from experience, that is, a carrying over of observed relations of uniform, or, as we may now also say, constant sequence to those which have not been or cannot be objects of observation, as well as to those which we expect to appear in the future. Psychologically regarded, it is merely the most general expression of an expectation, conditioned through associative reproduction, of uniform sequence. It is, therefore, — to bring Hume's doctrine to a conclusion that the father of modern empiricism himself did not draw, — a species of temporal contiguity.

The general validity which we ascribe to the causal law is accordingly a merely empirical one. It can never attain apodeictic or even assertorical validity, but purely that type of problematic validity which we may call "real" in contradistinction to the other type of problematic validity attained in judgments of objective as well as of subjective and hypothetical possibility.[1] No possible progress of experience can win for the empiristically interpreted causal law any other than this real problematic validity; for experience can never become complete *a parte post*, nor has it ever been complete *a parte ante*. The causal law is valid assertorically only in so far as it sums up, purely in the way of an inventory, the preceding experiences. We call such assumptions, drawn from well-ordered experience and of inductive origin, "hypotheses," whether they rest upon generalizing inductive inferences in the narrower sense, or upon specializing inferences from analogy. They, and at the same time the empiristically interpreted causal law, are not hypotheses in the sense in which Newton rightly rejected all formation of hypotheses,[2] but are such as are necessarily part of all methods in the sciences of facts in so far as the paths of research lead out beyond the content given immediately in perception to objects of only possible experience.

The assertion of Stuart Mill, in opposition to this conclusion, that the cause must be thought of as the "invariable antecedent" and, correspondingly, the effect as the "invariable consequent,"[3] does all honor to the genius of the thinker; but it agrees by no means with the empiristic presuppositions which serve as the basis for his conclusions. For, starting from these presuppositions, the "invariable sequence" can only mean one that is uniform and constant

[1] Cf. the author's *Logik*, bd. I, § 61.
[2] "*Rationem* vero harum gravitatis proprietatum ex phaenomenis nondum potui deducere, et hypotheses non fingo. *Quicquid enim ex phaenomenis non deducitur, hypothesis vocanda est;* et hypotheses seu metaphysicae, seu phvsicae, seu qualitatum occultarum, seu mechanicae, in philosophia experimentali locum non habent. In hac philosophia propositiones deducuntur ex phaenomenis, et redduntur generales per *inductionem*." Newton, at the end of his chief work.
[3] *Logic*, bk. III, ch. v, § 2.

according to past experience, and that we henceforth carry over to not yet observed events as far as these prove in conformity with it, and in this way verify the anticipation contained in our general assertion. The same holds of the assertion through which Mill endeavors to meet the above-mentioned objection of Reid, namely, that the unchanging sequence must at the same time be demonstrably an "unconditional" one. The language in which experience speaks to us knows the term "the unconditioned" as little as the term "the unchangeable," even though this have, as Mill explains, the meaning that the effect "will be, whatever supposition we may make in regard to all other things," or that the sequence will "be subject to no other than negative conditions." For in these determinations there does not lie exclusively, according to Mill, a probable prediction of the future. "It is *necessary* to our using the word cause, that we should believe not only that the antecedent always *has* been followed by the consequent, but that as long as the present constitution of things endures, it always *will* be so." Likewise, Mill, the man of research, not the empiristic logician, asserts that there belongs to the causal law, besides this generality referring to all possible events of uniform sequence, also an "undoubted assurance;" although he could have here referred to a casual remark of Hume.[1] Such an undoubted assurance, "that for every event . . . there is a law to be found, if we only know where to find it," evidently does not know of a knowledge referred exclusively to experience.

Hence, if the causal law is, as empiricism to be consistent must maintain, only a general hypothesis which is necessarily subject to verification as experience progresses, then it is not impossible that in the course of experience events will appear that are not preceded or followed uniformly by others, and that accordingly cannot be regarded as causes or effects. According to this interpretation of the causal law, such exceptional events, whether in individual or in repeated cases of perception, must be just as possible as those which in the course of preceding experience have proved themselves to be members of series of constant sequence. On the basis of previous experience, we should only have the right to say that such exceptional cases are less probable; and we might from the same ground expect that, if they could be surely determined, they would only have to be regarded as exceptions to the rule and not, possibly, as signs of a misunderstood universal non-uniformity of occurrence. No one wants to maintain an empirical necessity, that is, a statement that so comprehends a present experience or an hypothesis developed

---

[1] *Logic*, bk. III, ch. v, § 6, and end of § 2. Hume says in a note to section VI of his *Enquiry concerning Human Understanding :* "We ought to divide arguments into *demonstrations, proofs, and probabilities.* By proofs meaning such arguments from experience as leave no room for doubt or opposition." The note stands in evident contrast to the well-known remarks at the beginning of section IV, pt. I.

on the basis of present experience that its contradictory is rationally impossible. An event preceded by no other immediately and uniformly as cause would, according to traditional usage, arise out of nothing. An event that was followed immediately and constantly by no other would accordingly be an event that remained without effect, and, did it pass away, it must disappear into nothing. The old thought, well known in its scholastic formulation, *ex nihilo nihil fit, in nihilum nihil potest reverti*, is only another expression for the causal law as we have interpreted it above. The contradictories to each of the clauses of the thought just formulated, that something can arise out of nothing and pass into nothing, remain therefore, as a consequence of empiricism, an improbable thought, to be sure, but none the less a thought to which a real possibility must be ascribed.

It was in all probability this that Stuart Mill wished to convey in the much-debated passage: "I am convinced that any one accustomed to abstraction and analysis, who will fairly exert his faculties for the purpose, will, when his imagination has once learnt to entertain the notion, find no difficulty in conceiving that in some one, for instance, of the many firmaments into which sidereal astronomy now divides the universe, events may succeed one another at random without any fixed law; nor can anything in our experience, or in our mental nature, constitute a sufficient, or indeed any, reason for believing that this is nowhere the case." For Mill immediately calls our attention to the following: "Were we to suppose (what it is perfectly possible to imagine) that the present order of the universe were brought to an end, and that a chaos succeeded in which there was no fixed succession of events, and the past gave no assurance of the future; if a human being were miraculously kept alive to witness this change, he surely would soon cease to believe in any uniformity. the uniformity itself no longer existing." [1]

We can throw light from another side upon the thought that lies in this outcome of the empiristic interpretation of the causal law. If we still desire to give the name "effect" to an event that is preceded uniformly by no other, and that we therefore have to regard as arising out of nothing, then we must say that it is the effect of itself, that is, its cause lies in its own reality, in short, that it is *causa sui*. Therefore the assumption that a *causa sui* has just as much real possibility as have the causes of our experience which are followed uniformly by another event, is a necessary consequence of the empiristic view of causation. This much only remains sure, there is nothing contained in our previous experience that in any way assures us of the validity of this possible theory.

The empiristic doctrine of causation requires, however, still fur-

[1] *Logic*, bk. iii, ch. xxi, § 1.

ther conclusions. Our scientific, no less than our practical thought has always been accustomed to regard the relation between cause and effect not as a matter of mere sequence, not therefore as a mere formal temporal one. Rather it has always, in both forms of our thought, stood for a *real* relation, that is, for a relation of *dynamic dependence* of effect upon cause. Accordingly, the effect *arises out of the cause, is engendered through* it, or *brought forth by* it.

The historical development of this dynamic conception of cause is well known. The old anthropopathic interpretation, which interpolates anthropomorphic and yet superhuman intervention between the events that follow one another uniformly, has maintained itself on into the modern metaphysical hypotheses. It remains standing wherever God is assumed as the first cause for the interaction between parts of reality. It is made obscure, but not eliminated, when, in other conceptions of the world, impersonal nature, fate, necessity, the absolute identity, or an abstraction related to these, appears in the place of God. On the other hand, it comes out clearly wherever these two tendencies of thought unite themselves in an anthropopathic pantheism. That is, it rests only upon a difference in strength between the governing religious and scientific interests, whether or not the All-One which unfolds itself in the interconnection and content of reality is thought of more as the immanent God, or more as substance. Finally, we do not change our position, if the absolute, self-active being (in all these theories a first cause is presupposed as *causa sui*) is degraded to a non-intellectual will.

However, the dynamic interpretation of cause has not remained confined to the field of these general speculations, just because it commanded that field so early. There is a second branch, likewise early evolved from the stem of the anthropopathic interpretation, the doctrine that the causal relations of dependence are effected through "forces." These forces adhere to, or dwell in, the ultimate physical elements which are thought of as masses. Again, as spiritual forces they belong to the "soul," which in turn is thought of as a substance. In the modern contrast between attractive and repulsive forces, there lies a remnant of the Empedoklean opposition between Love and Hate. In the various old and new hylozoistic tendencies, the concepts of force and its correlate, mass, are eclectically united. In consistent materialism as well as spiritualism, and in the abstract dynamism of energetics, the one member is robbed of its independence or even rejected in favor of the other.[1]

[1] Alongside of these dynamic theories, there are to be found mechanical ones that arose just as early and from the same source, viz., the practical *Weltanschauung*. It is not part of our purpose to discuss them. Their first scientific expression is to be found in the doctrine of effluences and pores in Empedokles and in Atomism.

It is evident in what light all these dynamic conceptions appear, when looked at from the standpoint of consistent extreme empiricism. These "forces," to consider here only this one of the dynamic hypotheses, help to explain nothing. The physical forces, or those which give rise to movement, are evidently not given to us as contents of sense perception, and at the most they can be deduced as non-sensuous foundations, not as contents of possible sense perception. The often and variously expressed belief that self perception reveals to us here what our senses leave hidden has proved itself to be in all its forms a delusion. The forces whose existence we assume have then an intuitable content only in so far as they get it through the uniformities present in repeated perceptions, which uniformities are to be "explained" through them. But right here their assumption proves itself to be not only superfluous but even misleading; for it makes us believe that we have offered an explanation, whereas in reality we have simply duplicated the given by means of a fiction, quite after the fashion of the Platonic doctrine of ideas. This endeavor to give the formal temporal relations between events, which we interpret as causes and effects, a dynamic real substructure, shows itself thus to be worthless in its contributions to our thought. The same holds true of every other dynamic hypothesis. The critique called forth by these contributions establishes therefore only the validity of the empiristic interpretation.

If, however, we have once come so far, we may not hold ourselves back from the final step. Empiricism has long ago taken this step, and the most consistent among its modern German representatives has aroused anew the impulses that make it necessary. Indeed, if we start from the empiristic presuppositions, we must recognize that there lies not only in the assumption of forces, but even in the habit of speaking of causes and effects, "a clear trace of fetishism." We are not then surprised when the statement is made: The natural science of the future, and accordingly science in general, will, it is to be hoped, set aside these concepts also on account of their formal obscurity. For, so it is explained, repetitions of like cases in which *a* is always connected with *b*, namely, in which like results are found under like circumstances, in short, the essence of the connection of cause and effect, exists only in the abstraction that is necessary to enable us to repicture the facts. In nature itself there are no causes and effects. *Die Natur ist nur einmal da.*

It is, again, Stuart Mill, the man of research, not the empiricist, that opposes this conclusion, and indeed opposes it in the form that Auguste Comte had given it in connection with thoughts that can be read into Hume's doctrine. Comte's "objection to the *word* cause is a mere matter of nomenclature, in which, as a matter of nomenclature, I consider him to be entirely wrong. . . . By reject-

59

ing this form of expression, M. Comte leaves himself without any term for marking a distinction which, however incorrectly expressed, is not only real, but is one of the fundamental distinctions in science." [1]

For my own part, the right seems to be on the side of Comte and his recent followers in showing the old nomenclature to be worn out, if viewed from the standpoint of empiricism. If the relation between cause and effect consists alone in the uniformity of sequence which is hypothetically warranted by experience, then it can be only misleading to employ words for the members of this purely formal relation that necessarily have a strong tang of real dynamic dependence. In fact, they give the connection in question a peculiarity that, according to consistent empiricism, it does not possess. The question at issue in the empiristically interpreted causal relation is a formal functional one, which is not essentially different, as Ernst Mach incidentally acknowledges, from the interdependence of the sides and angles of a triangle.

Here two extremes meet. Spinoza, the most consistent of the dogmatic rationalists, finds himself compelled in his formulation of the analytic interpretation of the causal relation handed down to him to transform it into a mathematical one. Mach, the most consistent of recent German empiricists, finds himself compelled to recognize that the empirically synthetic relation between cause and effect includes no other form of dependence than that which is present in the functional mathematical relations. (In Germany empiricism steeped in natural science has supplanted the naïve materialism saturated with natural science.) That the mathematical relations must likewise be subjected to a purely empirical interpretation, which even Hume denied them, is a matter of course.

However, this agreement of two opposing views is no proof that empiricism is on the right road. The empiristic conclusions to which we have given our attention do not succeed in defining adequately the specific nature of the causal relation; on the contrary, they compel us to deny such a relation. Thus they cast aside the concept that we have endeavored to define, that is, the judgment in which we have to comprehend whatever is peculiar to the causal connection. But one does not untie a knot by denying that it exists.

It follows from this self-destruction of the empiristic causal hypothesis that an additional element of thought must be contained in the relation of cause and effect besides the elements of reproductive recognition and those of identification and discrimination, all of which are involved in the abstract comprehension of uniform sequence. The characteristics of the causal connection revealed by our previous analysis constitute the necessary and perhaps adequate conditions for combining the several factual perceptions into the

[1] *Logic*, bk. III, ch. v, § 6.

abstract registering idea of uniform sequence. We may, therefore, expect to find that the element sought for lies in the tendency to extend the demand for causal connections over the entire field of possible experience; and perhaps we may at the same time arrive at the condition which led Hume and Mill to recognize the complete universality of the causal law in spite of the exclusively empirical content that they had ascribed to it. In this further analysis also we have to draw from the nature of our thought itself the means of guiding our investigation.

In the first place, all thought has a formal necessity which reveals itself in the general causal law no less than in every individual thought process, that is, in every valid judgment. The meaning of this formal necessity of thought is easily determined. If we presuppose, for example, that I recognize a surface which lies before me as green, then the perception judgment, "This surface is green," that is, the apprehension of the present perceptive content in the fundamental form of discursive thought, repeats with predicative necessity that which is presented to me in the content of perception. The necessity of thought contained in this perception judgment, as *mutatis mutandis* in every affirmative judgment meeting the logical conditions, is recognizable through the fact that the contradictory judgment, "This surface is not green," is impossible for our thought under the presupposition of the given content of perception and of our nomenclature. It contradicts itself. I can express the contradictory proposition, for instance, in order to deceive; but I cannot really pass the judgment that is contained in it. It lies in the very nature of our thought that the predicate of an assertive judgment can contain only whatever belongs as an element of some sort (characteristic, attribute, state, relation) to the subject content in the wider sense. The same formal necessity of thought, to give a further instance, is present in the thought process of mediate syllogistic predication. The conclusion follows necessarily from the premises, for example, the judgment, "All bodies are divisible," from the propositions, "All bodies are extended," and, "Whatever is extended is divisible."

These elementary remarks are not superfluous; for they make clear that the casually expressed assertion of modern natural scientific empiricism, declaring in effect that there is no such thing as necessity of thought, goes altogether too far. Such necessity can have an admissible meaning only in so far as it denotes that in predicting or recounting *the content* of possible experience every hypothesis is possible for thought. Of course it is, but that is not the subject under discussion.

The recognition of the formal necessity of thought that must be presupposed helps us to define our present question; for it needs no proof that this formal necessity of thought, being valid for every affirmative judgment, is valid also for each particular induction, and again for the general causal law. If in the course of our perceptions we meet uniform sequences, then the judgment, "These sequences are uniform," comprehends the common content of many judgments with formal necessity of thought. Empiricism, too, does not seriously doubt that the hypothesis of a general functional, even though only temporal, relation between cause and effect is deduced as an expectation of possible experience with necessity from our real experience. It questions only the doctrine that the relation between the events regarded as cause and effect has any other than a purely empirical import. The reality of an event that is preceded and followed uniformly by no other remains for this view, as we have seen, a possibility of thought.

In opposition to empiricism, we now formulate the thesis to be established: Wherever two events a and b are known to follow one another uniformly and immediately, there we must require with formal necessity that some element in the preceding a be thought of as fundamental, which will determine sufficiently b's appearance or make that appearance necessary. The necessity of the relation between the events regarded as cause and effect is, therefore, the question at issue.

We must keep in mind from the very start that less is asserted in this formulation than we are apt to read into it. It states merely that something in a must be thought of as fundamental, which makes b necessary. On the other hand, it says nothing as to what this fundamental something is, or how it is constituted. It leaves entirely undecided whether or not this something that our thought must necessarily postulate is a possible content of perception or can become such, accordingly whether or not it can become an object of our knowledge, or whether or not it lies beyond the bounds of all our possible experience and hence all our possible knowledge. It contains nothing whatsoever that tells us how the determination of b takes place through a. The word "fundamental" is intended to express all this absence of determination.

Thus we hope to show a necessity of thought peculiar to the relation between cause and effect. This is the same as saying that our proof will establish the logical impossibility of the contradictory assertion; for the logical impossibility of the contradictory assertion is the only criterion of logical necessity. Thus the proof that we seek can be given only indirectly. In the course of this proof, we can disregard the immediacy of the constant sequence and confine our attention to the uniformity of the sequence, not only for the sake of brevity, but also because, as we have seen, we have the right to speak of near and remote causes. We may then proceed as follows.

If there is not something fundamental in a constant antecedent event a, which determines necessarily the constant subsequent appearance of one and the same b, — that is, if there is nothing fundamental which makes this appearance necessary, — then we must assume that also c or d . . . , in short, any event you will, we dare not say "follows upon," but appears after a in irregular alternation with b. This assumption, however, is impossible for our thought, because it is in contradiction with our experience, on the basis of which our causal thought has been developed. Therefore the assumption of a something that is fundamental in a, and that determines sufficiently and necessarily the appearance of b, is a necessity for our thought.

The assertion of this logical impossibility (*Denkunmöglichkeit*) will at once appear thoroughly paradoxical. The reader, merely recalling the results of the empiristic interpretation given above, will immediately say: "The assumption that a b does not follow constantly upon an a, but that sometimes b, sometimes c, sometimes d . . . irregularly appears, is in contradiction only with all our previous experience, but it is not on this account a *logical* impossibility. It is merely improbable." The reader will appeal especially to the discussion of Stuart Mill, already quoted, in which Mill pictures *in concreto* such an improbable logical impossibility, and therefore at the same time establishes it in fact. Again, the reader may bring forward the words in which Helmholtz introduces intellectual beings of only two dimensions. "By the much misused expression, 'to be able to imagine to one's self,' or, 'to think how something happens,' I understand (and I do not see how anybody can understand anything else thereby without robbing the expression of all meaning) that one can picture to one's self the series of sense impressions which one would have if such a thing actually took place in an individual case."[1]

Nevertheless, pertinent as are these and similar objections, they are not able to stand the test. We ask: "Is in fact a world, or even a portion of our world, possible for thought that displays through an absolutely irregular alternation of events a chaos in the full sense; or is the attempt to picture such a chaos only a mere play of words to which not even our imagination, not to mention our thought, can give a possible meaning?"

Perhaps we shall reach a conclusion by the easiest way, if we subject Mill's description to a test. If we reduce it to the several propositions it contains, we get the following: (1) Every one is able to picture to himself in his imagination a reality in which events follow one another without rule, that is, so that after an event a now b appears, now c, etc., in complete irregularity. (2) The idea of

[1] *Vorträge und Reden*, bd. II, "Über den Ursprung und die Bedeutung der geometrischen Axiome."

such a chaos accordingly contradicts neither the nature of our mind nor our experience. (3) Neither the former nor the latter gives us sufficient reason to believe that such an irregular alternation does not actually exist somewhere in the observable world. (4) If such a chaos should be presented to us as fact, that is, if we were in a position to outlive such an alternation, then the belief in the uniformity of time relations would soon cease.

Every one would subscribe to the last of these four theses, immediately upon such a chaos being admitted to be a possibility of thought; that is, he would unless he shared the rationalistic conviction that our thought constitutes an activity absolutely independent of all experience. We must simply accept this conclusion on the ground of the previous discussion and of a point still to be brought forward.

If we grant this conclusion, however, then it follows, on the ground of our previous demonstration of the reproductive and recognitive, as well as thought elements involved in the uniform sequence, that the irregularity in the appearance of the events, assumed in such a chaos, can bring about an absolutely relationless alternation of impressions for the subject that we should presuppose to be doing the perceiving. If we still wish to call it perception, it would remain only a perception in which no component of its content could be related to the others, a perception, therefore, in which not even the synthesis of the several perception contents could be apprehended as such. That is, every combination of the different perception contents, by which they become components of one and the same perception, presupposes, as we have seen, those reproductive and recognitive acts in revival which are possible only where uniformities of succession (and of coexistence) exist. Again, every act of attention involved in identifying and discriminating, which likewise we have seen to be possible only if we presuppose uniformities in the given contents of perception, must necessarily disappear when we presuppose the chaotic content; and yet they remain essential to the very idea of such a chaos. A relationless chaos is after all nothing else than a system of relations thought of without relations! That the same contradiction obtains also in the mere mental picturing of a manifold of chaotic impressions needs no discussion; for the productive imagination as well as the reproductive is no less dependent than is our perceptive knowledge upon the reproductive recognition and upon the processes of identifying and discriminating.

Thus the mental image of a chaos could be formed only through an extended process of ideation, which itself presupposes as active in it all that must be denied through the very nature of the image. A relationless knowledge, a relationless abstraction, a relationless reproduction or recognition, a relationless identification or discrimination, in short, a relationless thought, are, as phrases, one and all mere contradictions. We cannot picture "through our relating thought," to use Helmholtz's expression, nor even in our imagination, the sense impressions that we should have if our thought were relationless, that is, were nullified in its very components and presuppositions. In the case of Helmholtz's two dimensional beings, the question at issue was not regarding the setting aside of the conditions of our thought and the substituting conditions contradictory to them, but regarding the setting aside of a part of the content of our sense intuition, meanwhile retaining the conditions and forms peculiar to our thought. In this case, therefore, we have a permissible fiction, whereas in Mill's chaos we have an unthinkable thought.

Again, the sense impressions that must be presupposed in an inherently relationless chaos have no possible relation to the world of our perception, whose components are universally related to each other through the uniformities of their coexistences and sequences. Accordingly, the remark with which Helmholtz concludes the passage above quoted holds, *mutatis mutandis*, here also. " If there is no sense impression known that stands in relation to an event which has never been observed (by us), as would be the case for us were there a motion toward a fourth dimension, and for those two dimensional beings were there a motion toward our third dimension; then it follows that such an 'idea' is impossible, as much so as that a man completely blind from childhood should be able to 'imagine' the colors, if we could give him too a conceptual description of them."

Hence the first of the theses in which we summed up Stuart Mill's assumptions must be rejected. With it go also the second and third. In this case we need not answer the question: In how far do these theses correspond to Mill's own statements regarding the absolute surety and universality of the causal law?

We have now found what we sought, in order to establish as a valid assertion the seeming paradox in the proof of the necessity that we ascribe to the relation between cause and effect. We have proved that the assumption of a completely irregular and therefore relationless alternation of impressions contradicts not only our experience, but even the conditions of our thought; for these presuppose the uniformities of the impressions, and consequently our ability to relate them, all which was eliminated from our hypothetical chaos. Hence we have also established that a necessary relation is implied in the thought of a constant sequence of events, which makes the uniformly following b really dependent upon the uniformly preceding a.

From still another side, we can make clear the necessity asserted in the relation of cause and effect. We found that the connection between each definite cause and its effect is an empirically synthetic one and has as its warrant merely experience. We saw further that the necessity inherent in the causal connection contains merely the demand that there shall be something fundamental in the constantly preceding a which makes the appearance of b necessary; not, however, that it informs us what this efficacy really is, and hence also not that it informs us how this efficacy brings about its effect. Finally, we had to urge that every induction, the most general no less than the most particular, depends upon the presupposition that the same causes will be given in the reality not yet observed as in that already observed. This expectation is warranted by no necessity of thought, not even by that involved in the relation of cause and effect; for this relation begins for future experience only when the presupposition that the same causes will be found in it is assumed as fulfilled.[1] This expectation is then dependent solely upon previous experience, whose servants we are, whose lords we can never be. Therefore, every induction is an hypothesis requiring the verification of a broader experience, since, in its work of widening and completing our knowledge, it leads us beyond the given experience to a possible one. In this respect we can call all inductive thought empirical, that is, thought that begins with experience, is directed to experience, and in its results is referred to experience. The office of this progressing empirical thought is accordingly to form hypotheses from which the data of perception can be regressively deduced, and by means of which they can be exhibited as cases of known relations of our well-ordered experience, and thus can be explained.

The way of forming hypotheses can be divided logically into different sections which can readily be made clear by an example. The police magistrate finds a human corpse under circumstances that eliminate the possibility of accident, natural death, or suicide; in short, that indicate an act of violence on the part of another man. The general hypothesis that he has here to do with a crime against life forms the guide of his investigation. The result of the circumstantial evidence, which we presuppose as necessary, furnishes then a special hypothesis as following from the general hypothesis.

It is clear that this division holds for all cases of forming hypotheses. A general hypothesis serves every special hypothesis as a heuristic principle. In the former we comprehend the causal explanation indicated immediately by the facts revealed to our perception

---

[1] The only empiricism which can maintain that the same causes would, in conformity with the causal law, be given in the unobserved reality, is one which puts all events that can be regarded as causes in the immediately given content of perception as its members. Such a view is not to be found in Mill; and it stands so completely in the way of all further analysis required of us by every perception of events that no attention has been paid in the text to this extreme of extremes.

in the special case. It contains, as we might also express it, the genus to the specific limitations of the more exact investigation. But each of these general hypotheses is a modification of the most general form of building hypotheses, which we have already come to know as the condition of the validity of all inductive inferences, that is, as the condition for the necessity of their deduction, and, consequently, as the condition for the thought that like causes will be given in the reality not yet observed as in that already observed. We have further noticed that in this most general form of building hypotheses there lie two distinct and different valid assumptions: beside the empirical statement that like causes will be given, which gives the inductive conclusion the hypothetical form, there stands the judgment that like causes bring forth like effects, a corollary of the causal law. The real dependence of the effect upon the cause, presupposed by this second proposition and the underlying causal law, is not, as was the other assumption, an hypothesis, but a necessary requirement or *postulate* of our thought. Its necessity arises out of our thought, because our experience reveals uniformity in the sequence of events. From this point of view, therefore, the causal law appears as a postulate of our thought, grounded upon the uniformity in the sequence of events. It underlies every special case of constructing hypotheses as well as the expectation that like causes will be given in the reality not yet observed.

Mill's logic of induction contains the same fault as that already present in Hume's psychological theory of cause. Hume makes merely the causal law itself responsible for our inductive inferences, and accordingly (as Mill likewise wrongly assumes) for our inferences in general. But we recognize how rightly Mill came to assert, in contradiction to his empiristic presuppositions, that the causal law offers "an undoubted assurance of an invariable, universal, and unconditional," that is, necessary, sequence of events, from which no seeming irregularity of occurrence and no gap in our experience can lead us astray, as long as experience offers uniformities of sequence.

Rationalism is thus in the right, when it regards the necessary connection as an essential characteristic of the relation between cause and effect, that is, recognizes in it a relation of real dependence. At this point Kant and Schopenhauer have had a profounder insight than Hume and Stuart Mill. Especially am I glad to be in agreement with Lotze on a point which he reached by a different route and from essentially different presuppositions. Lotze distinguishes in pure logic between postulates, hypotheses, and fictions. He does not refer the term "postulate" exclusively to the causal law which governs our entire empirical thought in its formation of hypotheses, but gives the term a wider meaning. "Postulates" are only corollaries from the inductive fundamental form of all hypothesis construction, and correspond essentially to what we have called general or heuristic hypotheses. His determination of the validity of these postulates, however, implies the position to be assigned to the causal law and therefore not to those heuristic hypotheses. " The postulate is not an assumption that we can make or refrain from making, or, again, in whose place we can substitute another. It is rather an (absolutely) necessary assumption without which the content of the view at issue would contradict the laws of our thought." [1]

Still the decision that we have reached is not on this account in favor of rationalism, as this is represented for instance by Kant and his successors down to our own time, and professed by Lotze in the passage quoted, when he speaks of an absolute necessity for thought. We found that the causal law requires a necessary connection between events given us in constant sequence. It is not, however, on that account a law of our thought or of a "pure understanding" which would be absolutely independent of all experience. When we take into consideration the evolution of the organic world of which we are members, then we must say that our intellect, that is, our ideation and with it our sense perception, has evolved in us in accordance with the influences to which we have been subjected. The common elements in the different contents of perception which have arisen out of other psychical elements, seemingly first in the brute world, are not only an occasion, but also an efficient cause, for the evolution of our processes of reproduction, in which our memory and imagination as well as our knowledge and thought, psychologically considered, come to pass. The causal law, which the critical analysis of the material-scientific methods shows to be a fundamental condition of empirical thought, in its requirement that the events stand as causes and effects in necessary connection, or real dependence, comprehends these uniform contents of perception only in the way peculiar to our thought.

Doubtless our thought gives a connection to experience through this its requirement which experience of itself could not offer. The necessary connection of effect with cause, or the real dependence of the former upon the latter, is not a component of possible perception. This requirement of our thought does not, however, become thereby independent of the perceptive elements in the presuppositions involved in the uniformity of sequence. The *a priori* in the sense of "innate ideas," denoting either these themselves or an absolutely *a priori* conformity to law that underlies them, for instance, our "spontaneity," presupposes in principle that our "soul" is an independently existing substance in the traditional metaphysical sense down to the time of Locke. Kant's rationalistic successors,

[1]*Logic*, 1874, buch II, kap. viii.

for the most part, lost sight of the fact that Kant had retained these old metaphysical assumptions in his interpretation of the transcendental conditions of empirical interaction and in his cosmological doctrine of freedom. The common root of the sensibility and of the understanding as the higher faculty of knowledge remains for Kant the substantial force of the soul, which expresses itself (just as in Leibnitz) as *vis passiva* and *vis activa*. The modern doctrine of evolution has entirely removed the foundation from this rationalism which had been undermined ever since Locke's criticism of the traditional concept of substance.

To refer again briefly to a second point in which the foregoing results differ from the Kantian rationalism as well as from empiricism since Hume: The postulate of a necessary connection between cause and effect, as we have seen, in no way implies the consequence that the several inductions lose the character of hypotheses. This does not follow merely from the fact that all inductions besides the causal law include the hypothetical thought that the same causes will be given in the reality not yet observed as appear in that already observed. The hypothetical character of all inductive inferences is rather revealed through the circumstance that in the causal postulate absolutely nothing is contained regarding *what* the efficacy in the causes is, and *how* this efficacy arises.

Only such consequences of the foregoing interpretation of the causal law and of its position as one of the bases of all scientific construction of hypotheses may be pointed out, in conclusion, as will help to make easier the understanding of the interpretation itself.

The requirement of a necessary connection, or dependence, is added by our thought to the reproductive and recognitive presuppositions that are contained in the uniformity of the sequence of events. If this necessary connection be taken objectively, then it reveals as its correlate the requirement of a real dependence of effect upon cause. We come not only upon often and variously used rationalistic thoughts, but also upon old and unchangeable components of all empirical scientific thought, when we give the name "force" to the efficacy that underlies causes. The old postulate of a dynamic intermediary between the events that follow one another constantly retains for us, therefore, its proper meaning. We admit without hesitation that the word "force" suggests fetishism more than do the words "cause" and "effect;" but we do not see how this can to any degree be used as a counter-argument. All words that were coined in the olden time to express thoughts of the practical *Weltanschauung* have an archaic tang. Likewise all of our science and the greater part of our nomenclature have arisen out of the sphere of thought contained in the practical *Weltanschauung*, which centred early in fetishism and related thoughts. If, then, we try to free our scientific terminology from such words, we must seek refuge in the Utopia of a *lingua universalis*, in short, we must endeavor to speak a language which would make science a secret of the few. Or will any one seriously maintain that a thought which belongs to an ancient sphere of mental life must be false for the very reason that it is ancient?

In any case, it is fitting that we define more closely the sense in which we are to regard forces as the dynamic intermediaries of uniform occurrence. Force cannot be given as a content of perception either through our senses or through our consciousness of self; in the case of the former, not in our kinesthetic sensations, in the case of the latter, not in our consciousness of volition. Volition would not include a consciousness of force, even though we were justified in regarding it as a simple primitive psychosis, and were not compelled rather to regard it as an intricate collection of feelings and sensations as far as these elementary forms of consciousness are connected in thought with the phenomena of reaction. Again, forces cannot be taken as objects that are derived as *possible* perceptions or after the analogy of possible perceptions. The postulate of our thought through which these forces are derived from the facts of the uniform sequence of events, reveals them as limiting notions (*Grenzbegriffe*), as specializations of the necessary connection between cause and effect, or of the real dependence of the former upon the latter; for the manner of their causal intermediation is in no way given, rather they can be thought of only as underlying our perceptions. They are then in fact *qualitates occultae;* but they are such only because the concept of quality is taken from the contents of our sense and self perception, which of course do not contain the necessary connection required by our thought. Whoever, therefore, requires from the introduction of forces new contents of perception, for instance, new and fuller mechanical pictures, expects the impossible.

The contempt with which the assumption of forces meets, on the part of those who make this demand, is accordingly easily understood, and still more easily is it understood, if one takes into consideration what confusion of concepts has arisen through the use of the term "force" and what obstacles the assumption of forces has put in the way of the material sciences. It must be frankly admitted that this concept delayed for centuries both in the natural and moral sciences the necessary analysis of the complicated phenomena forming our data. Under the influence of the "concept philosophy" it caused, over and over again, the setting aside of the problems of this analytical empirical thought as soon as their solution had been begun. This misuse cannot but make suspicious from the very start every new form of maintaining that forces underlie causation.

However, misuse proves as little here against a proper use as it does in other cases. Moreover, the scruples that we found arising from the standpoint of empiricism against the assumption of forces are not to the point. In assuming a dynamic intermediary between cause and effect, we are not doubling the problems whose solution is incumbent upon the sciences of facts, and still less is it true that our assumption must lead to a logical circle. That is, a comparison with the ideas of the old concept philosophy, which even in the Aristotelian doctrine contain such a duplication, is not to the point. Those ideas are hypostasized abstractions which are taken from the uniformly coexisting characteristics of objects. Forces, on the other hand, are the imperceivable relations of dependence which we must presuppose between events that follow one another uniformly, if the uniformity of this sequence is to become for us either thinkable or conceivable. The problems of material scientific research are not doubled by this presupposition of a real dynamic dependence, because it introduces an element not contained in the data of perception which give these problems their point of departure. This presupposition does not renew the thought of an analytic rational connection between cause and effect which the concept philosophy involves; on the contrary, it remains true to the principle made practical by Hume and Kant, that the real connection between causes and their effects is determinable only through experience, that is, empirically and synthetically through the actual indication of the events of uniform sequence. How these forces are constituted and work, we cannot know, since our knowledge is confined to the material of perception from which as a basis presentation has developed into thought. The insight that we have won from the limiting notion of force helps us rather to avoid the misuse which has been made of the concept of force. A fatal circle first arises, when we use the unknowable forces and not the knowable events for the purpose of explanation, that is, when we cut off short the empirical analysis which leads *ad infinitum*. To explain does not mean to deduce the known from the unknown, but the particular from the general. It was therefore no arbitrary judgment, but an impulse conditioned by the very nature of our experience and of our thought, that made man early regard the causal connection as a dynamic one, even though his conception was of course indistinct and mixed with confusing additions.

The concept of force remains indispensable also for natural scientific thought. It is involved with the causal law in every attempt to form an hypothesis, and accordingly it is already present in every description of facts which goes by means of memory or abstraction beyond the immediately given content of present perception. In introducing it we have in mind, moreover, that the foundations of every possible interpretation of nature possess a dynamic character, just because all empirical thought, in this field as well, is subordinate to the causal law. This must be admitted by any one who assumes as indispensable aids of natural science the mechanical figures through which we reduce the events of sense perception to the motion of mass particles, that is, through which we associate these events with the elements of our visual and tactual perception. All formulations of the concept of mass, even when they are made so formal as in the definition given by Heinrich Hertz, indicate dynamic interpretations. Whether the impelling forces are to be thought of in particular as forces acting at a distance or as forces acting through collision depends upon the answer to the question whether we have to assume the dynamic mass particles as filling space discontinuously or continuously. The dynamic basis of our interpretation of nature will be seen at once by any one who is of the opinion that we can make the connection of events intelligible without the aid of mechanical figures, for instance, in terms of energy.

Thus it results that we interpret the events following one another immediately and uniformly as causes and effects, by presupposing as fundamental to them forces that are the necessary means of their uniformity of connection. What we call "laws" are the judgments in which we formulate these causal connections.

A second and a third consequence need only be mentioned here. The hypothesis that interprets the mutual connection of psychical and physical vital phenomena as a causal one is as old as it is natural. It is natural, because even simple observations assure us that the mental content of perception *follows* uniformly the instigating physical stimulus and the muscular movement the instigating mental content which we apprehend as will. We know, however, that the physical events which, in raising the biological problem, we have to set beside the psychical, do not take place in the periphery of our nervous system and in our muscles, but in the central nervous system. But we must assume, in accordance with all the psycho-physiological data which at the present time are at our disposal, that these events in our central nervous system do not follow the corresponding psychical events, but that both series have their course simultaneously. We have here, therefore, instead of the real relation of dependence involved in constant sequence, a real dependence of the simultaneity or correlative series of events. This would not, of course, as should be at once remarked, tell as such against a causal connection between the two separate causal series. But the contested parallelistic interpretation of this dependence is made far more probable through other grounds. These are in part corollaries of the law of the conservation of energy, rightly interpreted, and in part epistemological considerations. Still it is not advisable to burden methodological study, for instance, the theory of induction, with these remote problems; and on that account it is better for our present investigation to subordinate the psychological interdependences to the causal ones in the narrower sense.

The final consequence, too, that forces itself upon our attention is close at hand in the preceding discussion. The tradition prevailing since Hume, together with its inherent opposition to the interpretation of causal connection given by the concept philosophy, permitted us to make the uniform sequences of events the basis of our discussion. In so doing, however, our attention had to be called repeatedly to one reservation. In fact, only a moment ago, in alluding to the psychological interdependences, we had to emphasize the uniform *sequence*. Elsewhere the arguments depended upon the *uniformity* that characterizes this sequence; and rightly, for the reduction of the causal relation to the fundamental relation of the sequence of events is merely a convenient one and not the only possible one. As soon as we regard the causal connection, along with the opposed and equal reaction, as an interconnection, then cause and effect become, as a matter of principle, simultaneous. The separation of interaction from causation is not justifiable.

In other ways also we can so transform every causal relation that cause and effect must be regarded as simultaneous. Every stage, for instance, of the warming of a stone by the heat of the sun, or of the treaty conferences of two states, presents an effect that is simultaneous with the totality of the acting causes. The analysis of a cause that was at first grasped as a whole into the multiplicity of its constituent causes and the comprehension of the constituent causes into a whole, which then presents itself as the effect, is a necessary condition of such a type of investigation. This conception, which is present already in Hobbes, but especially in Herbart's "method of relations," deserves preference always where the purpose in view is not the shortest possible argumentation but the most exact analysis.

If we turn our attention to this way of viewing the problem, — not, however, in the form of Herbart's speculative method, — we shall find that the results which we have gained will in no respect be altered. We do, however, get a view beyond. From it we can find the way to subordinate not only the uniform sequence of events, but also the persistent characteristics and states with their mutual relations, under the extended causal law. In so doing, we do not fall back again into the intellectual world of the concept philosophy. We come only to regard the *persisting coexistences* — in the physical field, the bodies, in the psychical, the subjects of consciousness — as systems or modes of activity. The thoughts to which such a doctrine leads are accordingly not new or unheard of. The substances have always been regarded as sources of modes of activity. We have here merely new modifications of thoughts that have been variously developed, not only from the side of empiricism, but also from that of rationalism. They carry with them methodologically the implication that it is possible to grasp the totality of reality, as far as it reveals uniformities, as a causally connected whole, as a cosmos. They give the research of the special sciences the conceptual bases for the wider prospects that the sciences of facts have through hard labor won for themselves. The subject of consciousness is unitary as far as the processes of memory extend, but it is not simple. On the contrary, it is most intricately put together out of psychical complexes, themselves intricate and out of their relations; all of which impress upon us, psychologically and, in their mechanical correlates, physiologically, an ever-recurring need for further empirical analysis. Among the mechanical images of physical reality that form the foundation of our interpretation of nature, there can finally be but one that meets all the requirements of a general hypothesis of the continuity of kinetic connections. With this must be universally coördinated the persistent properties or sensible modes of action belonging to bodies. The mechanical constitution of the compound bodies, no matter at what stage of combination and formation, must be derivable from the mechanical constitution of the elements of this combination. Thus our causal thought compels us to trace back the persistent coexistences of the so-called elements to combinations whose analysis, as yet hardly begun, leads us on likewise to indefinitely manifold problems. Epistemologically we come finally to a universal phenomenological dynamism as the fundamental basis of all theoretical interpretation of the world, at least fundamental for our scientific thought, and we are here concerned with no other.

Administration Building, Washington University

CONGRESS OF
ARTS AND SCIENCE

UNIVERSAL EXPOSITION, ST. LOUIS, 1904

EDITED BY

HOWARD J. ROGERS, A.M., LL.D.

DIRECTOR OF CONGRESSES

VOLUME IV

PHYSICS
CHEMISTRY
ASTRONOMY
SCIENCES OF THE EARTH

BOSTON AND NEW YORK
HOUGHTON, MIFFLIN AND COMPANY
The Riverside Press, Cambridge
1906

Robert Simpson Woodward
1848–1924

DIVISION C—PHYSICAL SCIENCE

(*Hall 4, September 20, 10 a. m.*)

Speaker: Professor Robert S. Woodward, Columbia University.

THE UNITY OF PHYSICAL SCIENCE

BY ROBERT SIMPSON WOODWARD

[Robert Simpson Woodward, Ph.D., Sc.D., LL.D., President of the Carnegie Institution of Washington. b. Rochester, Mich., 1849. C.E. University of Michigan, 1872; Ph.D. University of Michigan, 1892; Honorary LL.D. University of Wisconsin, 1904; Sc.D., University of Pennsylvania, and Columbia University, 1905. Assistant engineer, U. S. Lake Survey, 1872–82; assistant astronomer, U. S. Transit of Venus Commission, 1882–84; astronomer, geographer, and chief geographer, U. S. Geological Survey, 1884–90; assistant, U. S. Coast and Geodetic Survey, 1890–93; Professor of Mechanics and Mathematical Physics, Columbia University, 1893–1905; Dean of School of Pure Science, *ibid.*, 1895–1905; President of Carnegie Institution of Washington, 1905. Member of National Academy of Sciences; Past President and Treasurer (since 1894) of American Association for the Advancement of Science; Past President of American Mathematical Society and of New York Academy of Sciences; member of Astronomical and Astrophysical Society of America, Geological Society of America, Physical Society of America, and Washington Academy of Sciences. Author of *Smithsonian Geographical Tables ; Higher Mathematics* (with Mansfield Merriman); also of many Government reports and numerous papers and addresses on subjects in astronomy, geodesy, mathematics, mathematical physics, and education.]

There is a tradition, still tacitly sanctioned even by men of science, that there have been epochs when the more eminent minds were able to compass the entire range of knowledge. Amongst the vanishing heroic figures of the past it seems possible, indeed, to discern, here and there, a Galileo, a Huygens, a Descartes, a Leibnitz, a Newton, a Laplace, or a Humboldt, each capable, at least, of summing up with great completeness the state of contemporary knowledge. Traditions, however, are generally more or less mythical, and the myth in this case seems to be in flat contradiction with the fact that there never was such an epoch, that the great masters of our distinguished predecessors were, after all, much like the masters of to-day, simply the leading specialists of their times. But however this may be, if we grant the possibility of the requisite attainments, even in a few individuals at any epoch, we shall speedily conclude that there never was an epoch so much in need of them as the immediate present, when the divisional speakers of this Congress are called upon to explain the unities which pervade the ever-widening and largely diverse fields of their several domains.

The domain of physical science, concerning which I have the honor to address you to-day, presents peculiar and peculiarly formidable difficulties in the way of a summary review. While we may not be disposed to limit the wide range of inclusion specified by our programme, we must at once disclaim any attempt to speak authoritatively with respect to most of its details. There is, in fact, such a vast array of knowledge now comprehended under any one of the six Departments of our Division, that the boldest author must hesitate to enter on a limited discussion with respect to any of them. But if it is thus difficult to consider any department of physical science, it appears incomparably more difficult to contemplate all of them in the bewildering complexity of their interrelations and in the bewildering diversity of their subject-matter. What, for example, could seem more appalling to the average man of science than the duty of explaining the connections of archeology and astrophysics, or those of ecology and electrons?

Happily, however, the managers of the Congress have provided an adequate division of labor, whereby the technical details of the various Departments are allotted to experts, giving thus to a divisional speaker a degree of freedom with respect to depth in some way commensurate with the breadth of his task. Presuming, therefore, that I may deal only with the broader outlines and salient features of the subject, I invite your attention to a summary view of the present status and the apparent trend of physical science.

Whatever may be affirmed with respect to science in general, there appears to be no doubt that all of the physical sciences are characterized by three remarkable unities, — a unity of origin, a unity of growth, and a unity of purpose. Physical science originates in observation and experiment; it rises from the fact-gathering stage of unrelated qualities to the higher plane of related quantities, and passes thence on to the realm of correlation, computation, and prediction under theory; and its purpose is to interpret in consistent and verifiable terms the universe, of which we form a part. The recognition of these unities is of prime importance; for it helps us to understand and to anticipate a great diversity of perfection amongst the different branches of science, and hence leads us to appreciate the desirability of hearty coöperation on the part of scientific workers in order that progress may be ever positive towards the common goal.

Glancing rapidly *seriatim* at the different departments of physical science as specified by our programme, we come first to a consideration of formal physics, and we may most quickly orient ourselves aright in this department by trying to state in what respects the physics of to-day differs from the physics of a hundred years ago.

In spite of the extraordinary perfection of the work of Lagrange, Laplace, Fourier, Young, Fresnel, Poisson, Green, Gauss, and others of the early part of the nineteenth century, it will be at once admitted that great progress has been made. In addition to noteworthy advances and improvements along the lines laid down by these masters, there have been developed the relatively new fields of elasticity, electromagnetics, thermodynamics, and astrophysics; and there has been discovered the widest of all generalizations in physical science, — the law of conservation of energy. Whereas it was easy a century ago to conceive, as in gravitational astronomy, of action at a distance across empty space, the universe in the mean time has come to appear more and more plethoric not only with "gross matter," but with that most wonderful entity we call the ether. The astronomers have shown us, in fact, that the number of molar systems in the universe is enormously greater than was supposed possible a century ago; while the physicists have revealed to us molecular systems rivaling our solar system and its Jovian and Saturnian subsystems, and they have loaded down the ether with a burden of properties and relationships which its usual tenuity seems scarcely fitted to bear. Whereas, also, a century ago the tendency of thought, under the stimulus of the remarkable developments of the elastic solid theory of light and the fluid theories of electricity, was chiefly towards an ether whose continuity would have pleased Anaxagoras, the tendency to-day is chiefly towards an ether whose atomicity would have pleased Democritus.

On the whole, it must be said that the advances of the past century, and especially those of the past half-century, have been mainly along the lines of molecular physics. The epoch of Laplace was distinctly an epoch of molar physics; the epoch of to-day is distinctly an epoch of molecular physics. Light, heat, electricity, and magnetism have been definitely correlated as molecular and ethereal phenomena; while the recently discovered X-rays and the wonders of radioactivity, along with the "electrons," the "corpuscles" and the "electrions" of current investigations, all point towards a molecular constitution of the ether. Thermodynamics, likewise, large as it has grown in recent decades, is essentially a development of the molecular theory of gases. It would be too bold, perhaps, to assert that the trend of accumulating knowledge is towards an atomic unity of matter, but the day seems not far distant when there will be room for a new *Principia* and for a treatise which will accomplish for molecular systems what the *Mécanique Céleste* accomplished for the solar system.

One of the most important advances of recent decades is found in the fixation of ideas with respect to the units of physical science, and in the great improvements which have been wrought in metrology by the "International Bureau of Weights and Measures." Our standards of length, mass, and time are now fixed with a degree of precision which leaves little to be desired for the present; and the capital resources of measurement and calculation are now available to an extent never hitherto approached.

It should be noted, however, that confidence in the stability of our standards is by no means comparable with the perfection of their current applications. Indeed, we may raise with respect to them the question so long mooted with regard to the motions of the members of the solar system: namely, are they stable? Notwithstanding the admirable precision of the intercomparisons of the prototype meters and prototype kilograms and the equally admirable precision of Professor Michelson's determination of the length of the meter in terms of wave-lengths of cadmium light, we cannot affirm that these observed relations will hold indefinitely. Our inherited notions of mass have been rather rudely shaken, also, by the penetrating criticisms of Mach, and it appears possible even that the law of conservation of mass may need modification in the light of pending researches. But worst of all, our time-unit, the sidereal day, is so far from possessing the element of constancy that we may affirm with practical certainty that it is secularly variable. Having realized, through Professor Michelson's superb determination just referred to, the cosmic standard of length suggested by Maxwell thirty years ago, we are now much more in need of an equally trustworthy cosmic standard of time.

If the progress of physics during the past century has been chiefly in the direction of atomic theory, the progress of chemistry has been still more so. Chemistry is, in fact, the science of atoms and molecules *par excellence*, a distinction it has maintained for well-nigh a full century under the dominance of the fruitful atomic and molecular hypotheses of Dalton and of Avogadro and Ampère, and under the similarly fruitful laws of gases established by Dalton and Gay-Lussac. Perhaps the most striking feature of this progress, in a general way, is the gradual disappearance it has entailed of the imaginary lines which have been long thought to separate the fields of chemistry and physics. Through the remarkable discoveries of Faraday the two fields have been found to overlap in actual electrical contact. Through the wonderful revelations of spectrum analysis, originating with Bunsen and Kirchhoff, they have been proved to be very largely common ground. And through the broader generalizations inaugurated by Willard Gibbs, Helmholtz, and others, they are now both somewhat in danger of being annexed as a sub-province of rational mechanics.

To one whose work has fallen more especially in the fields of precise astronomy, geodesy, or metrology, it might seem a just reproach to chemistry that it is a science whose measurements and calculations demand, as a rule, no greater arithmetical resources than those of four-place tables of logarithms and anti-logarithms. The so-called "Constants of Nature" supplied by chemistry are, in fact, known with a low degree of certainty; a degree expressed, say, by three to five significant figures. A small amount of reflection, however, will convince one that the phenomena with which the chemist has to deal are usually far more complex than those which have yielded the splendid precision of astronomy, geodesy, and metrology. Moreover, it should be observed that the certainties even of these highly perfected sciences are very unequal in their different branches. It appears more correct, therefore, as well as more just, considering the central position it occupies and the wide range of its ramifications, along with the vast aggregate of qualitative and quantitative knowledge it has massed, to assert that the precision of chemistry affords the best numerical index of the present state of physical science. That is, when reduced to the most compact form of statement, the certainties of physical science are best indicated, in a general way, by a table of the combining weights of the eighty-odd chemical elements.

When one contemplates the numbers of such a table, and when one adds to its suggestions those which flow from the various periodic groupings of the same numbers, he can hardly avoid being inspired by the day-dreams of those who have looked long for the atomic unity of matter. But however the grand problem which thus obtrudes itself may be resolved finally, it appears certain that this table must stand as one of the great landmarks along the path of progress in physical science.

It was justly remarked by Laplace in his *Système du Monde* that "L'Astronomie, par la dignité de son objet et par la perfection de ses théories, est le plus beau monument de l'esprit humain, le titre le plus noble de son intelligence " ; and we must all admit that subsequent progress has gone far to maintain this high position for the most ancient and interesting of the older sciences. One finds little difficulty in accounting for the early rise of astronomical science and for the universal interest in celestial phenomena. Their immanence and omnipresence appeal even to the dullest intellects. But it is not so easy to account for the remarkable fact that although astronomy deals chiefly with the relations of bodies separated by immense distances, progress in its development has thus far been at least equal to, if not in advance of, the progress of physics and chemistry, which have to deal with matter close at hand. Without attempting a full explanation of this fact, it may suffice to observe that the principal phenomena of astronomy thus far developed appear to be relatively simple in comparison with those of the other physical sciences; and that the immense distances which separate the celestial bodies, instead of being an obstacle to, are a fortunate circumstance directly in favor of, the triumphant advances which have distinguished astronomical science from the epoch of Galileo down to the present day.

Not less noteworthy than his high estimate of the position of astronomy in his time are Laplace's anticipations of the course of future progress. Our admiration is kindled by the clearness of his vision with respect to ways and means, and by the penetration of his predictions of future discoveries. Advances in sidereal astronomy, he rightly thought, would depend chiefly on improvements in telescopes; while advances in dynamical astronomy were to come along with increased precision in the observed places of the members of the solar system and along with the growing perfection of analysis. It is almost needless to say that Laplace's brilliant anticipations have been quite surpassed by the actual developments. Observational astronomy has become one of the most delicately perfect of all the sciences; dynamical astronomy easily outstrips all competitors in the perfection of its theories and in the certainty of its predictions; while the newly developed branch of astrophysics supplies the last link in the chain of evidence of the essential unity of the material universe.

The order of the dimensions and the order of the mass contents of the visible universe, at any rate, have been pretty clearly made out. In addition to the vast aggregate of direct observational evidence collected and recorded during the past century, numerous theoretical researches have gone far, also, to interpret the laws which reign in the apparent chaos of the stars. The solar system, with its magnificent subsystems, has been proved to exhibit the type of stellar systems in general.

In a profound investigation recently published, Lord Kelvin has sought to correlate under the law of gravitation the principal observed data of the visible universe. Assuming this universe to lie within a sphere of radius equal to the distance of a star whose parallax is one thousandth of a second of arc, he concludes that there must be something like a thousand million masses of the magnitude of our sun within that sphere. Light traveling at the rate of 300,000 kilometers per second would require about six thousand years to traverse the diameter of this universe, and while the average distance asunder of the visible stars is considerably less, it is still of the same order. It is only essential, therefore, to imagine our luminary surrounded by a thousand million such suns, most of which are, in all probability, attended by groups of planets, to get some idea of the quantity of matter within visual range of our relatively insignificant terrestrial abode. And the imposing range of the astronomer's time-scale is perhaps impressively brought home to us when we reflect that a million years is the smallest convenient unit for recording the life-history of a star, while the current events in that history are transmitted across the interstellar medium by vibrations which occur at the rate of about six hundred million million times per second. Measured by its accumulation of achievements, then, the astronomy of to-day fulfills the requirements of a highly developed science. It is characterized by a vast aggregate of accurately determined facts related by theories founded on a small number of hypotheses. In the past it has called forth the two greatest of all systematic treatises, the *Principia* of Newton and the *Mécanique Céleste* of Laplace. It has probably done more also than any other science, up to the present time, to illuminate the dark periods during which man has floundered in his struggle for advancement; and the indications are that its prestige will long continue.

But there are spots on every sun; and lest some may infer, even humorously, as Carlyle did seventy-odd years ago, that our system of the world is "as good as perfect," attention should be called to some noteworthy defects in astronomical data and to some singular obscurities in astronomical theory. Here, however, great caution and brevity are essential to avoid poaching on the preserves of our colleagues of the Sections. It may suffice, therefore, merely to mention, under the head of defective data, the low precision of the solar parallax, the aberration constant, the masses of the members of the solar system, and the uncertainty of our time-unit, already referred to. Two instances, likewise, which belong to the general field of physics as well, may suffice as illustrations of obscurities in astronomical theory. Stated in the order of their apparent complexity, these obscurities refer to the law of gravitation and to the phenomenon of stellar aberration. Probably both are related, and one may hope that any explanation of either will throw light on the other.

So long as no attempt is made to reconcile the law of gravitation with other branches of physics, progress, up to a certain point, is easy; and probably great advantage has resulted from the fact that dynamical astronomers have not been seriously disturbed by a desire to harmonize this law with the more elementary laws of mechanics. Perhaps they have unconsciously rested on the platform that gravitation is one of the "primordial causes" which are impenetrable to us. There are some indications that even Laplace and Fourier did so rest. However this may be, it has grown steadily more and more imperative during the past century to explain gravitation, or to discover the mechanism which provides that the force between two widely separated masses is proportional to their product directly and to the square of the distance between them inversely. All evidence seems to indicate that the ether must provide this mechanism; but, strangely enough, so far, the ether has baffled all attempts to reveal the secret. The problem has been attacked also on the purely

observational side of the numerical value of the gravitation constant.
But the splendid experimental researches for this purpose throw no
light on the mechanism in question, and, unfortunately, they bring
out values for the constant of a low order of precision.

With regard to stellar aberration, it must be at once admitted that
we have neither an adequate theory nor a precisely determined fact.
The astronomer has generally contented himself with the elementary
view that aberration is a purely kinematical phenomenon; that the
earth not only slips through the ether without sensible retardation,
but that the ether slips through the earth without sensible effects.
This difficulty was recognized, in a way, by Young and Fresnel, and,
although the subject of elaborate investigation in recent decades, it
has proved equally baffling with Newtonian gravitation. As in the
case of the latter also, the numerous attempts made to determine
the constant of aberration by observational methods have been rewarded by results of only meagre precision. Possibly the time has
arrived when one may raise the question, Within what limits is it
proper to speak of a gravitation constant or of an aberration constant?

If we agree with Laplace that astronomy is entitled to the highest
rank among the physical sciences, we can accord nothing short of
second place to the sciences of the earth. Most of them are, indeed,
intimately related to astronomy; and some of them are scarcely
less ancient in their origins, less dignified in their objects, or less
perfect in their theories. Primarily, also, it should be observed, geophysics is not simply a part of, but is the very foundation of, astronomy; for the earth furnishes the orientation, the base-line, and the
timepiece by means of which the astronomer explores the heavens.
Geology, likewise, in the broader sense of the term, as we are now
coming to see, is a fundamental science not only by reason of its
interpretations of terrestrial phenomena, but also by reason of its
parallel interpretations of celestial phenomena; for there is little
doubt that in the evolution of the earth we may read a history which
is in large degree typical of the history of celestial bodies. In any
revised estimate, therefore, of the relative rank of the physical
sciences, while it would be impossible to lower the science of the
heavens, it would appear essential to raise the sciences of the earth
to a much higher plane of importance than was thought appropriate
by our predecessors of a hundred years ago.

As with physics, chemistry, and astronomy, the wonderful progress
of the nineteenth century in geophysical science has been along
lines converging towards the more recondite properties of matter.
All parts of the earth, through observation, experiment, induction,
and deduction, have yielded increasing evidence of limited unities
amid endless diversities. Adopting the convenient terminology of

geologists for the different shells of the earth, let us glance rapidly in turn at the sciences of the atmosphere, the hydrosphere or oceans, the lithosphere or crust, and the centrosphere or nucleus.

The atmosphere is the special province of meteorologists, and although they are not yet able to issue long-range predictions, like those guaranteed by our theories of tides and terrestrial magnetism, it must be admitted that they have made great progress towards a rational description of the apparently erratic phenomena of the weather. One of the peculiar anomalies of this science illustrates in a striking way the general need of additional knowledge of the properties of matter; in this case, especially, the properties of gases. It is the fact that in meteorology greater progress has been made, up to date, in the interpretation of the kinetic than in the interpretation of the static phenomena of the atmosphere. Considering that static properties are usually much simpler than kinetic properties, it seems strange that we should know much more about cyclones, for example, than we do about the mass and the mass distribution of the atmosphere. In respect to this apparently simple question meteorology seems to have made no advance beyond the work of Laplace. There are indications, however, that this, along with many other questions, must await the advent of a new *Principia*.

The geodesists, who are the closest allies of the astronomers, may be said to preside over the hydrosphere, since most of their theories as well as most of their observations are referred to the sea level. They have determined the shape and the size of the earth to a surprising degree of certainty; but they are now confronted by problems which depend chiefly on the mass and mass distribution of the earth. The exquisite refinement of their observational methods has brought to light a minute wandering in the earth of its axis of rotation, which makes the latitude of any place a variable quantity; but the interpretation of this phenomenon is again a physical and not a mensurational problem. They have worked improvements also in all kinds of apparatus for refined measurements, as of baselines, angles, and differences of level; but here, likewise, they appear to approach limits set by the properties of matter.

The lithosphere was once thought to be the restricted province of geologists, but they now lay claim to the entire earth, from the centre of the centrosphere to the limits of the atmosphere, and they threaten to invade the region of the astronomers on their way toward the outlying domain of cosmogony. Geology illustrates better than any other science, probably, the wide ramifications and the close interrelations of physical phenomena. There is scarcely a process, a product, or a principle in the whole range of physical science, from physics and chemistry up to astronomy and astrophysics, which is not fully illustrated in its uniqueness or in its diversity by actual operations still in progress on the earth, or by actual records preserved in her crust. The earth is thus at once the grandest of laboratories and the grandest of museums available to man.

Any summary statement, from a non-professional student, of the advances in geology during the past century, would be hopelessly inadequate. Such a task could be fitly undertaken only by an expert, or by a corps of them. But out of the impressive array of achievements of this science, two seem to be especially worthy of general attention. They are the essential determination of the properties and the rôle of the lithosphere, and the essential determination of the time-scale suitable for measuring the historical succession of terrestrial events. The lithosphere is the theatre of the principal activities, mechanical and biological, of our planet; and a million years is the smallest convenient unit for recording the march of those activities. When one considers the intellectual as well as the physical obstacles which had to be surmounted, and when one recalls the bitter controversies between the Neptunists and the Vulcanists and between the Catastrophists and the Uniformitarians, these achievements are seen to be amongst the most important in the annals of science.

The centrosphere is the *terra incognita* whose boundaries only are accessible to physical science. It is that part of the earth concerning which astronomers, geologists, and physicists have written much, but concerning which, alas! we are still in doubt. Where direct observation is unattainable, speculation is generally easy, but the exclusion of inappropriate hypotheses is, in such cases, generally difficult. Nevertheless, it may be affirmed that the range of possibilities for the state of the centrosphere has been sharply restricted during the past half-century. Whatever may have been the origin of our planet, whether it has evolved from nebular condensation or from meteoric accretion; and whatever may be the distribution of temperature within the earth's mass as a whole; it appears certain that pressure is the dominant factor within the nucleus. Pressure from above, supplied in hydrostatic measure by the plastic lithosphere, supplemented by internal pressure below, must determine, it would seem, within narrow limits, the actual distribution of density throughout the centrosphere, regardless of its material composition, of its effective rigidity, or of its potential liquidity. Here, however, we are extending the known properties of matter quite beyond the bounds of experience, or of present possible experiment; and we are again reminded of the unity of our needs by the diversity of our difficulties.

In his recently published autobiography, Herbert Spencer asserts that at the time of issue of his work on biology (1864) "not one person in ten or more knew the meaning of the word . . . and among those who knew it, few cared to know anything about the subject." That the attitude of the educated public towards biological science could have been thus indifferent, if not inimical, forty years ago, seems strange enough now even to those of us who have witnessed in part the scientific progress subsequent to that epoch. But this was a memorable epoch, marked by the advent of the great intellectual awakening ushered in by the generalizations of Darwin, Wallace, Spencer, and their coadjutors. And the quarter of a century which immediately followed this epoch appears, as we look back upon it, like an heroic age of scientific achievement. It was an age during which some men of science, and more men not of science, lost their heads temporarily, if not permanently; but it was also an age during which most men of science, and thinking people in general, moved forward at a rate quite without precedent in the history of human advancement. A new, and a greatly enlarged, view of the universe was introduced in the doctrine of evolution, advanced and opposed, alike vigorously, chiefly by reason of its biological applications and implications. Galileo, Newton, and Laplace had given us a system of the inorganic world; Darwin, Spencer, and their followers have foreshadowed a system which includes the organic world as well.

The astonishing progress of biology in recent times furnishes the most convincing evidence of the unity and the efficiency of the methods of physical science in the interpretation of natural phenomena. For the biologist has followed the same methods, with changes appropriate to his subject-matter only, as those found fruitful in astronomy, chemistry, and all the rest. And whatever may be the increased complexity of the organic over the inorganic world, or however high the factor of life may seem to raise the problems of biology above the plane of the other physical sciences, there has appeared no sufficient reason, as yet, to doubt either the validity or the adequacy of those methods.

Moreover, the interrelations of biology with chemistry and physics especially are yearly growing more and more extended and intimate through the rapidly expanding researches of bacteriology, physiology, and physiological chemistry, plant and animal pathology, and so on, up through cytology to the embryology of the higher forms of life. Through the problems of these researches also we are again brought face to face, sooner or later, with the problems of molecular science.

And finally, what may be said of anthropology, which is at once the most interesting and the most novel of the physical sciences, — interesting by reason of its subject-matter, novel by reason of its applications? Some of us, perhaps, might be inclined to demur from a classification which makes man, along with matter, a fit object

of investigation in physical science. Granted even that he is usually
a not altogether efficient thermodynamic engine, it may yet appear
that he is worthy of a separate category. Fortunately, however, it
is not a rule of physical science to demand immediate answers to
such ulterior questions. It is enough for the present to know that
man furnishes no exception, save in point of complexity, to the manifestations of physical phenomena so widely exhibited in the animal
kingdom.

But whatever may be our inherited prejudices, or our philosophic
judgments, we are confronted by the fact that the study of man
in all his attributes is now an established domain of science. And
herein we rise to a table-land of transcendent fascination; for, to
adapt a phrase of an eminent master in physical science, the instruments of investigation are the objects of research. Herein also we
find the culminating unity, not only of the physical sciences, but of
all of the sciences; and it is chiefly for the promotion of these higher
interests of anthropology that we are assembled in this cosmopoli-
tan congress to-day.

It has been our good fortune to witness in recent decades an unparalleled series of achievements in the fields of physical science.
All of them, from anthropology and astronomy up to zoölogy, have
yielded rich harvests of results; and one is prone to raise the question
whether a like degree of progress may be expected to prevail during
the century on which we have now entered. No man can tell what
a day may bring forth; much less may one forecast the progress of
a decade or a century. But, judging from the long experience of the
past, there are few reasons to doubt and many reasons to expect
that the future has still greater achievements available. It would
appear that we have found the right methods of investigation. Philosophically considered, the remarkable advances of the past afford
little cause for marvel. On the contrary, they are just such results
as we should anticipate from persistent pursuit of scientific investigation. Conscious of the adequacy of his methods, therefore, the
devotee to physical science has every inducement to continue his
labors with unflagging zeal and confident optimism.

DEPARTMENT IX — PHYSICS

(*Hall* 6, *September* 20, 2 *p. m.*)

CHAIRMAN: PROFESSOR HENRY CREW, Northwestern University.
SPEAKERS: PROFESSOR EDWARD L. NICHOLS, Cornell University.
PROFESSOR CARL BARUS, Brown University.

THE Chairman of the Department of Physics was Professor Henry Crew, of Northwestern University, who opened the proceedings of the Department by saying: " Whatever views we may entertain concerning the classification of the sciences which Professor Münsterberg has proposed for the guidance of this congress, we will, I believe, all concur in the opinion that it is full of suggestion and very instructive. For my own part, I think it gives a really profound glimpse into the relationships of the various departments of human learning. You will recall that the first main division is between the pure and applied sciences. We have come together this afternoon to consider a subject which lies in the former group. But physics is not the only pure science: it is merely one belonging to that subdivision which deals with phenomena. Again, there are two classes of phenomena, the mental and the physical: and physics has to do only with the latter class. Indeed, it does not cover the entire field of physical phenomena, but constitutes merely one of the six Departments in this Division. Physics is, however, the most general and most fundamental of this group of six. It is properly found, therefore, at the head of the list. Our theme this afternoon, then, is that fundamental science which deals with the general properties of matter and energy and which includes the general principles of all physical phenomena. We are fortunate in having with us men who, by wide experience gained in their own researches, and by a thorough study of the philosophy of the subject, are eminently fitted to treat this topic."

Edward L. Nichols
1854–1937

THE FUNDAMENTAL CONCEPTS OF PHYSICAL SCIENCE

BY EDWARD LEAMINGTON NICHOLS

[Edward Leamington Nichols, Professor of Physics, Cornell University, and Editor-in-chief of the *Physical Review*. b. September 14, 1854, Leamington, England. B.S. Cornell University, 1875; Ph.D. Göttingen, 1879; Fellowship in Physics, Johns Hopkins University, 1879–80; Professor of Physics and Chemistry, Central University, 1881–83; Professor of Physics and Astronomy, University of Kansas, 1883–87. Member of National Academy of Science, American Academy of Arts and Sciences, American Institute of Electrical Engineers, American Philosophical Society, American Physical Society. Author of *A Laboratory Manual of Physics* and *Applied Electricity; The Outlines of Physics*, etc.]

ALL algebra, as was pointed out by von Helmholtz [1] nearly fifty years ago, is based upon the three following very simple propositions:

Things equal to the same thing are equal to each other.

If equals be added to equals the wholes are equal.

If unequals be added to equals the wholes are unequal.

Geometry, he adds, is founded upon a few equally obvious and simple axioms.

The science of physics, similarly, has for its foundation three fundamental conceptions: those of *mass, distance,* and *time,* in terms of which all physical quantities may be expressed.

Physics, in so far as it is an exact science, deals with the relations of these so-called physical quantities; and this is true not merely of those portions of the science which are usually included under the head of physics, but also of that broader realm which consists of the entire group of the physical sciences, viz., astronomy, the physics of the heavens; chemistry, the physics of the atom; geology, the physics of the earth's crust; biology, the physics of matter imbued with life; physics proper (mechanics, heat, electricity, sound, and light).

The manner in which the three fundamental quantities L, M, and T (length, mass, and time) enter, in the case of a physical quantity, is given by its *dimensional formula*.

Thus the dimensional formula for an acceleration is LT^{-2} which expresses the fact that an acceleration is a velocity (a length divided by a time) divided by a time. Energy has for its dimensional formula L^2MT^{-2}; it is a force, $LT^{-2}M$ (an acceleration multiplied by a mass), multiplied by a distance.

Not all physical quantities, in the present state of our knowledge, can be assigned a definite dimensional formula, and this indicates that not all of physics has as yet been reduced to a clearly established

[1] Von Helmholtz, *Populäre Wissenschaftliche Vorträge*, p. 136.

mechanical basis. The dimensional formula thus affords a valuable criterion of the extent and boundaries of our strictly definite knowledge of physics. Within these boundaries we are on safe and easy ground, and are dealing, independent of all speculation, with the relations between precisely defined quantities. These relations are mathematical, and the entire superstructure is erected upon the three fundamental quantities, L, M, and T, and certain definitions; just as geometry arises from its axioms and definitions.

Of many of those physical quantities, for which we are not as yet able to give the dimensional formula, our knowledge is precise and definite, but it is incomplete. In the case, for example, of one important group of quantities, those used in electric and magnetic measurements, we have to introduce, in addition to L, M, and T, a constant factor to make the dimensional formula complete. This, the *suppressed factor* of Rücker,[1] is μ, the magnetic permeability, when the quantity is expressed in the electromagnetic system, and becomes k, the specific inductive capacity, when the quantity is expressed in terms of the electrostatic system.

Here the existence of the suppressed factor is indicative of our ignorance of the mechanics involved. If we knew in what way a medium like iron increased the magnetic field, or a medium like glass the electric field, we should probably be able to express μ and k in terms of the three selected fundamental dimensions and complete the dimensional formulæ of a large number of quantities.

Where direct mechanical knowledge ceases, the great realm of physical speculation begins. It is the object of such speculation to place all phenomena upon a mechanical basis; excluding as unscientific all occult, obscure, and mystical considerations.

Whenever the mechanism by means of which phenomena are produced is incapable of direct observation either because of its remoteness in space, as in the case of physical processes occurring in the stars, or in time, as in the case of the phenomena with which the geologist has to do, or because of the minuteness of the moving parts, as in molecular physics, physical chemistry, etc., the speculative element is unavoidable. Here we are compelled to make use of analogy. We infer the unknown from the known. Though our logic be without flaw, and we violate no mathematical principle, yet are our conclusions not absolute. They rest of necessity upon *assumptions*, and these are subject to modification indefinitely as our knowledge becomes more complete.

A striking instance of the uncertainties of extrapolation and of the precarious nature of scientific assumptions is afforded by the various estimates of the temperature of the sun. Pouillet placed this temperature between 1461°C. and 1761°C.; Secchi at 5,000,000°; Ericsson

[1] Rücker, *Philos. Mag.*, 27, p. 104. 1889.

at 2,500,000°. The newer determinations[1] of the temperature of the surface are, to be sure, in better agreement. Le Chatelier finds it to be 7600°; Paschen, 5400°; Warburg, 6000°. Wilson and Gray publish as their corrected result 8000°. The estimate of the internal temperature is of a more speculative character. Schuster's computation gives 6,000,000° to 15,000,000°; that of Kelvin, 200,000,000°; that of Ekholm, 5,000,000°.

Another interesting illustration of the dangers of extrapolation occurs in the history of electricity. Faraday, starting from data concerning the variation between the length of electric sparks through air with the difference of potential, made an interesting computation of the potential difference between earth and sky necessary to discharge a cloud at a height of one mile. He estimated the difference of potential to be about 1,000,000 volts. Later investigations of the sparking distance have, however, shown this function to possess a character quite different from that which might have been inferred from the earlier work, and it is likely that Faraday's value is scarcely nearer the truth than was the original estimate of the temperature of the sun, mentioned above.

Still another notable instance of the errors to which physical research is subject when the attempt is made to extend results beyond the limits established by actual observation occurs in the case of the measurements of the infra-red spectrum of the sun by Langley. His beautiful and ingenious device, the bolometer, made it possible to explore the spectrum to wave-lengths beyond those for which the law of dispersion of the rock-salt prism had at that time been experimentally determined. Within the limits of observation the dispersion showed a curve of simple form, tending apparently to become a straight line as the wave-length increased. There was nothing in the appearance of the curve to indicate that it differed in character from the numerous empirical curves of similar type employed in experimental physics, or to lead even the most experienced investigator to suspect values for the wave-length derived from an extension of the curve. The wave-lengths published by Langley were accordingly accepted as substantially correct by all other students of radiation; but subsequent measurements of the dispersion of rock salt at the hands of Rubens and his co-workers showed the existence of a second sudden and unlooked-for turn of the curve just beyond the point at which the earlier determinations ceased; and in consequence Langley's wave-lengths and all work based upon them are now known to be not even approximately accurate. The history of physics is full of such examples of the dangers of extrapolation, or, to speak more broadly, of the tentative character of most of our assumptions in experimental physics.

[1] See Arrhenius, *Kosmische Physik*, p. 131.

We have, then, two distinct sets of physical concepts. The first of these deals with that positive portion of physics, the mechanical basis of which, being established upon direct observation, is fixed and definite, and in which the relations are as absolute and certain as those of mathematics itself. Here speculation is excluded. Matter is simply one of the three factors, which enters, by virtue of its mass, into our formulæ for energy, momentum, etc. Force is simply a quantity of which we need to know only its magnitude, direction, point of application, and the time during which it is applied. The Newtonian conception of force — the producer of motion — is adequate. All troublesome questions as to how force acts, of the mechanism by means of which its effects are produced, are held in abeyance.

Speculative physics, to which the second set of concepts belongs, deals with those portions of the science for which the mechanical basis has to be imagined. Heat, light, electricity, and the science of the nature and ultimate properties of matter belong to this domain.

In the history of the theory of heat we find one of the earliest manifestations of a tendency so common in speculative physics that it may be considered characteristic: the assumption of a medium. The medium in this case was the so-called *imponderable* caloric; and it was one of a large class, of which the two electric fluids, the magnetic fluid, etc., were important members.

The theory of heat remained entirely speculative up to the time of the establishment of the mechanical equivalent of heat by Joule. The discovery that heat could be measured in terms of work injected into thermal theory the conception of energy, and led to the development of thermodynamics.

Generalizations of the sort expressed by Tyndall's phrase, *heat a mode of motion*, follow easily from the experimental evidence of the part which energy plays in thermal phenomena, but the specification of the precise mode of motion in question must always depend upon our views concerning the nature of matter, and can emerge from the speculative stage only, if ever, when our knowledge of the mechanics of the constitution of matter becomes fixed. The problem of the mechanism by which energy is stored or set free rests upon a similar speculative basis.

These are proper subjects for theoretical consideration, but the dictum of Rowland [1] that we get out of mathematical formulæ only what we put into them should never be lost from sight. So long as we put in only assumptions we shall take out hypotheses, and useful as these may prove, they are to be regarded as belonging to the realm of scientific speculation. They must be recognized as subject to modification indefinitely as we, in consequence of increasing knowledge, are led to modify our assumptions.

[1] Rowland, *President's Address to the American Physical Society*, 1900.

The conditions with which the physicist has to deal in his study of optics are especially favorable to the development of the scientific imagination, and it is in this field that some of the most remarkable instances of successful speculative work are to be found. The emission theory died hard, and the early advocates of the undulatory theory of light were forced to work up, with a completeness probably without parallel in the history of science, the evidence, necessarily indirect, that in optics we have to do with a wave-motion. The standpoint of optical theory may be deemed conclusive, possibly final, so far as the general proposition is concerned that it is the science of a wave-motion. In a few cases, indeed, such as the photography of the actual nodes of a standing wave-system, by Wiener, we reach the firm ground of direct observation.

Optics has nevertheless certain distinctly speculative features. Wave-motion demands a medium. The enormous velocity of light excludes known forms of matter; the transmission of radiation *in vacuo* and through outer space from the most remote regions of the universe, and at the same time through solids such as glass, demands that this medium shall have properties very different from that of any substance with which chemistry has made us acquainted.

The assumption of a medium is, indeed, an intellectual necessity, and the attempt to specify definitely the properties which it must possess in order to fulfill the extraordinary functions assigned to it has afforded a field for the highest display of scientific acumen. While the problem of the mechanism of the luminiferous ether has not as yet met with a satisfactory solution, the ingenuity and imaginative power developed in the attack upon its difficulties command our admiration.

Happily the development of what may be termed the older optics did not depend upon any complete formulation of the mechanics of the ether. Just as the whole of the older mechanics was built up from Kepler's laws, Newton's laws of motion, the law of gravitational attraction, the law of inverse squares, etc., without any necessity of describing the mechanics of gravitation or of any force, or of matter itself, so the system of geometrical relations involved in the consideration of reflection and refraction, diffraction, interference, and polarization was brought to virtual completion without introducing the troublesome questions of the nature of the ether and the constitution of matter.

Underlying this field of geometrical optics, or what I have just termed the older optics, are, however, a host of fundamental questions of the utmost interest and importance, the treatment of which depends upon molecular mechanics and the mechanics of the ether. Our theories as to the nature and causes of radiation, of absorption, and of dispersion, for example, belong to the newer optics, and are based upon our conceptions of the constitution of matter; and since our ideas concerning the nature of matter, like our knowledge of the ether, is purely speculative, the science of optics has a doubly speculative basis. One type of selective absorption, for example, is ascribed to resonance of the particles of the absorbing substance, and our modern dispersion theories depend upon the assumption of natural periods of vibration of the particles of the refracting medium of the same order of frequency as that of the light-waves. When the frequency of the waves falling upon a substance coincides with the natural period of vibration of the particles of the latter, we have selective absorption, and accompanying it, anomalous dispersion. For these and numerous other phenomena no adequate theory is possible which does not have its foundation upon some assumed conception as to the constitution of matter.

The development of the modern idea of the ether forms one of the most interesting chapters in the history of physics. We find at first a tendency to assume a number of distinct media corresponding to the various effects (visual, chemical, thermal, phosphorescent, etc.) of light-waves, and later the growth of the conception of a single medium, the luminiferous ether.

In the development of electricity and magnetism, meantime, the assumption of media was found to be an essential — something without which no definite philosophy of the phenomena was possible. At first there was the same tendency to a multiplicity of media — there were the positive and negative electric fluids, the magnetic fluid, etc. Then there grew up in the fertile mind of Faraday that wonderful fabric of the scientific imagination, the electric field; the conception upon which all later attempts to form an idea of a thinkable mechanism of electric and magnetic action have been established.

It is the object of science, as has been pointed out by Ostwald, to reduce the number of hypotheses; the highest development would be that in which a single hypothesis served to elucidate the relations of the entire universe. Maxwell's discovery that the whole theory of optics is capable of expression in terms identical with those found most convenient and suitable in electricity, in a word, that optics may be treated simply as a branch of electromagnetics, was the first great step towards such a simplification of our fundamental conceptions. This was followed by Hertz's experimental demonstration of the existence of artificially produced electromagnetic waves in every respect identical with light-waves, an achievement which served to establish upon a sure foundation the conception of a single medium. The idea of one universal medium as the mechanical basis for all physical phenomena was not altogether new to the theoretical physicist, but the unification of optics and electricity did much to strengthen this conception.

The question of the ultimate structure of matter, as has already been pointed out, is also speculative in the sense that the mechanism upon which its properties are based is out of the range of direct observation. For the older chemistry and the older molecular physics the assumption of an absolutely simple atom and of molecules composed of comparatively simple groupings of such atoms sufficed. Physical chemistry and that new phase of molecular physics which has been termed the physics of the ion demand the breaking up of the atom into still smaller parts and the clothing of these with an electric charge. The extreme step in this direction is the suggestion of Larmor that the electron is a "disembodied charge" of negative electricity. Since, however, in the last analysis, the only conception having a definite and intelligible mechanical basis which physicists have been able to form of an electric charge is that which regards it as a phenomenon of the ether, this form of speculation is but a return under another name to views which had earlier proved attractive to some of the most brilliant minds in the world of science, such as Helmholtz and Kelvin. The idea of the atom, as a vortex motion of a perfect fluid (the ether), and similar speculative conceptions, whatever be the precise form of mechanism imagined, are of the same class as the moving electric charge of the later theorists.

Lodge,[1] in a recent article in which he attempts to voice in a popular way the views of this school of thought, says:

"Electricity under strain constitutes 'charge'; electricity in locomotion constitutes light. What electricity itself is we do not know, but it may, perhaps, be a form or aspect of matter. . . . Now we can go one step further and say, matter is composed of electricity and of nothing else. . . ."

If for the word *electricity* in this quotation from Lodge we substitute *ether*, we have a statement which conforms quite as well to the accepted theories of light and electricity as his original statement does to the newer ideas it is intended to express.

This reconstructed statement would read as follows:

Ether under strain constitutes "charge"; *ether* in locomotion constitutes current and magnetism; *ether* in vibration constitutes light. What *ether* itself is we do not know, but it may, perhaps, be a form or aspect of matter. Now we can go one step further and say: "Matter is composed of *ether* and of nothing else."

The use of the word electricity, as employed by Lodge and others, is now much in vogue, but it appears to me unfortunate. It would be distinctly conducive to clearness of thought and an avoidance of confusion to restrict the term to the only meaning which is free from criticism; that in which it is used to designate the science which deals with electrical phenomena.

[1] Lodge, *Harper's Magazine*, August, 1904, p. 383.

The only way in which the noun *electricity* enters, in any definite and legitimate manner, into our electrical treatises is in the designation of Q in the equations —

$$Q = \int I dt, \ C = Q|E, \ W = QE, \text{ etc.}$$

Here we are in the habit — whether by inheritance from the age of the electric fluid, by reason of the hydrodynamic analogy, or as a matter of convention or of convenience merely — of calling Q the *quantity of electricity.*

Now Q is "charge" and its unit, the coulomb, is unit-charge. The alternative expression, *quantity of electricity*, is a purely conventional designation and without independent physical significance. It owes its prevalence among electricians to the fact that by virtue of long familiarity we prefer to think in terms of matter, which is tangible, rather than of ether. Charge is to be regarded as fundamental, and its substitute, quantity of electricity, as merely an artificial term of convenience; because of the former we have a definite mechanical conception, whereas we can intelligently define a quantity of electricity only in terms of *charge.*

In the science of heat the case differs, in that the term heat is used, if not as precisely synonymous with energy, at least for a quantity having the same dimensions as energy and having as its unit the erg. It might easily have happened, as has happened in electrical theory, that the ancient notion of a *heat substance* should survive, in which case we should have had for the quantity of heat not something measured in terms of energy, but, as in the case of electricity, one of the terms which enter into our expression for energy. We should then have had to struggle continually, in thermodynamics, as we now do in electrical theory, against the tendency to revert to an antiquated and abandoned view.

It would, I cannot but think, have been fortunate had the word electricity been used for what we now call electrical energy; using *charge*, or some other convenient designation, for the quantity Q. That aspect of the science in accordance with which we regard it as a branch of energetics in which movements of the ether are primarily involved would have been duly emphasized. We should have been quit forever of the bad notion of electricity as a medium, just as we are already freed from the incubus of heat as a medium. We should have had *electricity — a mode of motion* (or stress), *ether*, as we have *heat — a mode of motion of matter*. When our friends asked us: "What is electricity?" we should have had a ready answer for them instead of a puzzled smile.

One real advance which has been attained by means of the theory of ionization, and it is of extreme significance and of far-reaching importance, consists in the discovery that electrification, or the possession of charge, instead of being a casual or accidental property, temporarily imparted by friction or other process, is a fundamental property of matter. According to this newer conception of matter, the fruit of the ionic theory, the ultimate parts of matter are electrically charged particles. In the language of Rutherford: [1]

"It must then be supposed that the process of ionization in gases consists in a removal of a negative corpuscle or electron from the molecule of gas. At atmospheric pressure this corpuscle immediately becomes the centre of an aggregation of molecules which moves with it and *is* the negative ion. After removal of the negative ion the molecule retains a positive charge and probably also becomes the centre of a cluster of new molecules.

"The *electron* or *corpuscle* is the body of smallest mass yet known to science. It carries a negative charge of 3.4×10^{-10} electrostatic units. Its presence has only been detected when in rapid motion, when it has for speeds up to about 10^{10} cms. a second, an apparent mass m given by $e/m - 1.86 \times 10^7$ electromagnetic units. This apparent mass increases with the speed as the velocity of light is approached."

At low pressures the electron appears to lose its load of clustering molecules, so that finally the negative ion becomes identical with the electron or corpuscle, and has a mass, according to the estimates of J. J. Thomson, about one thousandth of that of the hydrogen atom. The positive ion is, however, supposed to remain of atomic size even at low pressures.

The ionic theory and the related hypothesis of electrolytic dissociation afford a key to numerous phenomena concerning which no adequate or plausible theories had hitherto been formed. By means of them explanations have been found, for example, of such widely divergent matters as the positive electric charge known to exist in the upper atmosphere, and the perplexing phenomena of fluorescence.

The evidence obtained by J. J. Thomson and other students of ionization, that electrons from different substances are identical, has greatly strengthened the conviction which for a long time has been in process of formation in the minds of physicists, that all matter is in its ultimate nature identical. This conception, necessarily speculative, has been held in abeyance by the facts, regarded as established, and lying at the foundation of the accepted system of chemistry, of the conservation of matter and the intransmutability of the elements. The phenomena observed in recent investigations of radioactive substances have, however, begun to shake our faith in this principle.

If matter is to be regarded as a product of certain operations performed upon the ether, there is no theoretical difficulty about

[1] Rutherford, *Radioactivity*, p. 53. 1904.

transmutation of elements, variation of mass, or even the complete disappearance or creation of matter. The absence of such phenomena in our experience has been the real difficulty, and if the views of students of radioactivity concerning the transformations undergone by uranium, thorium, and radium are substantiated, the doctrines of the conservation of mass and matter which lie at the foundation of the science of chemistry will have to be modified. There has been talk of late of violations of the principle of the conservation of energy in connection with the phenomena of radioactivity, but the conservation of matter is far more likely to lose its place among our fundamental conceptions.

The development of physics on the speculative side has led, then, to the idea, gradually become more definite and fixed, of a universal medium, the existence of which is a matter of inference. To this medium properties have been assigned which are such as to enable us to form an intelligible, consistent conception of the mechanism by means of which phenomena, the mechanics of which is not capable of direct observation, may be logically considered to be produced. The great step in this speculation has been the discovery that a single medium may be made to serve for the numerous phenomena of optics, and that, without ascribing to it any characteristics incompatible with a luminiferous ether, it is equally available for the description and explanation of electric and magnetic fields, and finally may be made the basis for intelligible theories of the structure of matter.

To many minds this seemingly universal adaptability of the ether to the needs of physics almost removes it from the field of speculation; but it should not be forgotten that a system, entirely imaginary, may be devised, which fits all the known phenomena and appears to offer the only satisfactory explanation of the facts, and which subsequently is abandoned in favor of other views. The history of physics is full of instances where a theory is for a time regarded as final on account of its seeming completeness, only to give way to something entirely different.

In this consideration of the fundamental concepts I have attempted to distinguish between those which have the positive character of mathematical laws and which are entirely independent of all theories of the ultimate nature of matter, and those which deal with the latter questions and which are essentially speculative. I have purposely refrained from taking that further step which plunges us from the heights of physics into the depths of philosophy.

With the statement that science in the ultimate analysis is nothing more than *an attempt to classify and correlate our sensations* the physicist has no quarrel. It is, indeed, a wholesome discipline for him to formulate for himself his own relations to his science in terms such as those which, to paraphrase and translate very freely the opening passages of his recent *Treatise on Physics*, Chwolson [1] has employed.

"For every one there exist two worlds, an inner and an outer, and our senses are the medium of communication between the two. The outer world has the property of acting upon our senses, to bring about certain changes, or, as we say, to exert certain stimuli.

"The inner world, for any individual, consists of all those phenomena which are absolutely inaccessible (so far as direct observation goes) to other individuals. The stimulus from the outer world produces in our inner world a subjective perception which is dependent upon our *consciousness*. The subjective perception is made objective, viz., is assigned *time* and *place* in the outer world and given a name. The investigation of the processes by which this objectivication is performed is a function of philosophy."

Some such confession of faith is good for the man of science, — *lest he forget;* but once it is made he is free to turn his face to the light once more, thankful that the *investigation of objectivication* is, indeed, *a function of philosophy*, and that the only speculations in which he, as a physicist, is entitled to engage are those which are amenable at every step to mathematics and to the equally definite axioms and laws of mechanics.

---

[1] Chwolson, *Physik*, vol. I, Introduction.

Carl Barus
1856–1935

THE PROGRESS OF PHYSICS IN THE NINETEENTH CENTURY

BY CARL BARUS

[Carl Barus, Dean of the Graduate Department, Brown University. b. February 19, 1856, Cincinnati, Ohio. Ph.D. Columbia University, University of Würzburg, Bavaria. Physicist, U. S. Geological Survey; Professor of Meteorology, U. S. Weather Bureau; Professor of Physics, Smithsonian Institution; Member of the National Academy of Science of the United States; Vice-President of American Association for the Advancement of Science; Corresponding Member of the British Association for the Advancement of Science; Honorary Member of the Royal Institution of Great Britain; President of American Physical Society; Rumford Medalist. Author of *The Laws of Gases; The Physical Properties of the Iron Carburets;* and many other books; contributor to the standard magazines.]

You have honored me by requesting at my hands an account of the advances made in physics during the nineteenth century. I have endeavored, in so far as I have been able, to meet the grave responsibilities implied in your invitation; yet had I but thought of the overwhelmingly vast territory to be surveyed, I well might have hesitated to embark on so hazardous an undertaking. To mention merely the *names* of men whose efforts are linked with splendid accomplishments in the history of modern physics would far exceed the time allotted to this address. To bear solely on certain subjects, those, for instance, with which I am more familiar, would be to develop an unsymmetrical picture. As this is to be avoided, it will be necessary to present a straightforward compilation of all work above a certain somewhat vague and arbitrary lower limit of importance. Physics is, as a rule, making vigorous though partial progress along independent parallel lines of investigation, a discrimination between which is not possible until some cataclysm in the history of thought ushers in a new era. It will be essential to abstain from entering into either explanation or criticism, and to assume that all present are familiar with the details of the subjects to be treated. I can neither popularize nor can I endeavor to entertain, except in so far as a rapid review of the glorious conquests of the century may be stimulating.

In spite of all this simplicity of aim, there is bound to be distortion. In any brief account, the men working at the beginning of the century, when investigations were few and the principles evolved necessarily fundamental, will be given greater consideration than equally able and abler investigations near the close, when workers (let us be thankful) were many, and the subjects lengthening into detail. Again, the higher order of genius will usually be additionally exalted at the expense of the less gifted thinker. I can but regret that these are the inevitable limitations of the cursory treatment prescribed.

As time rolls on, the greatest names more and more fully absorb the activity of a whole epoch.

Metrology

Finally, it will hardly be possible to consider the great advances made in physics except on the theoretical side. Of renowned experimental researches, in particular of the investigations of the constants of nature to a degree of ever-increasing accuracy, it is not practicable to give any adequate account. Indeed, the refinement and precision now demanded have placed many subjects beyond the reach of individual experimental research, and have culminated in the establishment of the great national or international laboratories of investigation at Sèvres (1872), at Berlin (1887, 1890), at London (1900), at Washington (1901). The introduction of uniform international units in cases of the arts and sciences of more recent development is gradually, but inexorably, urging the same advantages on all. Finally, the access to adequate instruments of research has everywhere become an easier possibility for those duly qualified, and the institutions and academies which are systematically undertaking the distribution of the means of research are continually increasing in strength and in number.

Classification

In the present paper it will be advisable to follow the usual procedure in physics, taking in order the advances made in dynamics, acoustics, heat, light, and electricity. The plan pursued will, therefore, specifically consider the progress in elastics, crystallography, capillarity, solution, diffusion, dynamics, viscosity, hydrodynamics, acoustics; in thermometry, calorimetry, thermodynamics, kinetic theory, thermal radiation; in geometric optics, dispersion, photometry, fluorescence, photochemistry, interference, diffraction, polarization, optical media; in electrostatics, Volta contacts, Seebeck contacts, electrolysis, electric current, magnetism, electromagnetism, electrodynamics, induction, electric oscillation, electric field, radioactivity.

Surely this is too extensive a field for any one man! Few who are not physicists realize that each of these divisions has a splendid and voluminous history of development, its own heroes, its sublime classics, often culled from the activity of several hundred years. I repeat that few understand the unmitigatedly fundamental character, the scope, the vast and profound intellectual possessions, of pure physics; few think of it as the one science into which all other sciences must ultimately converge — or a separate representation would have been given to most of the great divisions which I have named.

Hence even if the literary references may be given in print with some fullness, it is impossible to refer verbally to more than the chief actors, and quite impossible to delineate sharply the real significance and the relations of what has been done. Moreover, the dates will in most instances have to be omitted from the reading. It has been my aim, however, to collect the greater papers in the history of physics, and the suggestion is implied that science would gain if by some august tribunal researches of commanding importance were formally canonized for the benefit of posterity.

Elastics

To begin with elasticity, whose development has been of such marked influence throughout the whole of physics, we note that the theory is virtually a creation of the nineteenth century. Antedating Thomas Young, who in 1807 gave to the subject the useful conception of a modulus, and who seems to have definitely recognized the shear, there were merely the experimental contribution of Galileo (1638), Hooke (1660), Mariotte (1680), the elastic curve of J. Bernoulli (1705), the elementary treatment of vibrating bars of Euler and Bernoulli (1742), and an attempted analysis of flexure and torsion by Coulomb (1776).

The establishment of a theory of elasticity on broad lines begins almost at a bound with Navier (1821), reasoning from a molecular hypothesis to the equation of elastic displacement and of elastic potential energy (1822–1827); yet this startling advance was destined to be soon discredited, in the light of the brilliant generalizations of Cauchy (1827). To him we owe the six component stresses and the six component strains, the stress quadric and the strain quadric, the reduction of the components to three principal stresses and three principal strains, the ellipsoids, and other of the indispensable conceptions of the present day. Cauchy reached his equations both by the molecular hypothesis and by an analysis of the oblique stress across an interface, — methods which predicate fifteen constants of elasticity in the most general case, reducing to but one in the case of isotropy. Contemporaneous with Cauchy's results are certain independent researches by Lamé and Clapeyron (1828) and by Poisson (1829).

Another independent and fundamental method in elastics was introduced by Green (1837), who took as his point of departure the potential energy of a conservative system in connection with the Lagrangian principle of virtual displacements. This method, which has been fruitful in the hands of Kelvin (1856), of Kirchhoff (1876), of Neumann (1885), leads to equations with twenty-one constants for the æolotropic medium reducing to two in the simplest case.

The wave-motion in an isotropic medium was first deduced by Poisson in 1828, showing the occurrence of longitudinal and transverse waves of different velocities; the general problem of wavemotion in æolotropic media, though treated by Green (1842), was attacked with requisite power by Blanchet (1840–1842) and by Christoffel (1877).

Poisson also treated the case of radial vibrations of a sphere (1828), a problem which, without this restriction, awaited the solutions of Jaerisch (1879) and of Lamb (1882). The theory of the free vibrations of solids, however, is a generalization due to Clebsch (1857–58, *Vorlesungen*, 1862).

Elasticity received a final phenomenal advance through the long-continued labors of de St. Venant (1839–55), which in the course of his editions of the work of Moigno, of Navier (1863), and of Clebsch (1864), effectually overhauled the whole subject. He was the first to assert adequately the fundamental importance of the shear. The profound researches of de St. Venant on the torsion of prisms and on the flexure of prisms appeared in their complete form in 1855 and 1856. In both cases the right sections of the stressed solids are shown to be curved, and the curvature is succinctly specified; in the former Coulomb's inadequate torsion formula is superseded, and in the latter flexural stress is reduced to a transverse force and a couple. But these mere statements convey no impression of the magnitude of the work.

Among other notable creations with a special bearing on the theory of elasticity there is only time to mention the invention and application of curvilinear coördinates by Lamé (1852); the reciprocal theorem of Betti (1872), applied by Cerruti (1882) to solids with a plane boundary — problems to which Lamé and Clapeyron (1828) and Boussinesq (1879–85) contributed by other methods; the case of the strained sphere studied by Lamé (1854) and others; Kirchhoff's flexed plate (1850); Rayleigh's treatment of the oscillations of systems of finite freedom (1873); the thermo-elastic equations of Duhamel (1838), of F. Neumann (1841), of Kelvin (1878); Kelvin's analogy of the torsion of prisms with the supposed rotation of an incompressible fluid within (1878); his splendid investigations (1863) of the dynamics of elastic spheroids and the geophysical applications to which they were put.

Finally, the battle royal of the molecular school following Navier, Poisson, Cauchy, and championed by de St. Venant, with the disciples of Green, headed by Kelvin and Kirchhoff, — the struggle of the fifteen constants with the twenty-one constants, in other words, — seems to have temporarily subsided with a victory for the latter through the researches of Voigt (1887–89).

Crystallography

Theoretical crystallography, approached by Steno (1669), but formally founded by Haüy (1781, *Traité*, 1801), has limited its development during the century to systematic classifications of form. Thus the thirty-two type sets of Hessel (1830) and of Bravais (1850) have expanded into the more extensive point series involving 230 types due to Jordan (1868), Sohncke (1876), Federow (1890), and Schoenfliess (1891). Physical theories of crystalline form have scarcely been unfolded.

Capillarity

Capillarity antedated the century in little more than the provisional, though brilliant, treatment due to Clairaut (1743). The theory arose in almost its present state of perfection in the great memoir of Laplace (1805), one of the most beautiful examples of the Newton-Boscovichian (1758) molecular dynamics. Capillary pressure was here shown to vary with the principal radii of curvature of the exposed surface, in an equation involving two constants, one dependent on the liquid only, the other doubly specific for the bodies in contact. Integrations for special conditions include the cases of tubes, plates, drops, contact angle, and similar instances. Gauss (1829), dissatisfied with Laplace's method, virtually reproduced the whole theory from a new basis, avoiding molecular forces in favor of Lagrangian displacements, while Poisson (1831) obtained Laplace's equations by actually accentuating the molecular hypothesis; but his demonstration has since been discredited. Young in 1805 explained capillary phenomena by postulating a constant surface tension, a method which has since been popularized by Maxwell (*Heat*, 1872).

With these magnificent theories propounded for guidance at the very threshold of the century, one is prepared to anticipate the wealth of experimental and detailed theoretical research which has been devoted to capillarity. Among these the fascinating monograph of Plateau (1873), in which the consequences of theory are tested by the behavior both of liquid lamellæ and by suspended masses, Savart's (1833), and particularly Rayleigh's, researches with jets (1879–83), Kelvin's ripples (1871), may be cited as typical. Of peculiar importance, quite apart from its meteorological bearing, is Kelvin's deduction (1870) of the interdependence of surface tension and vapor pressure when varying with the curvature of a droplet.

111

Diffusion

Diffusion was formally introduced into physics by Graham (1850). Fick (1855), appreciating the analogy of diffusion and heat conduction, placed the phenomenon on a satisfactory theoretical basis, and Fick's law has since been rigorously tested, in particular by H. F. Weber (1879).

The development of diffusion from a physical point of view followed Pfeffer's discovery (1877) of osmotic pressure, soon after to be interpreted by van 't Hoff (1887) in terms of Boyle's and Avogadro's laws. A molecular theory of diffusion was thereupon given by Nernst (1887).

Dynamics

In pure dynamics the nineteenth century inherited from the eighteenth that unrivaled feat of reasoning called by Lagrange the *Mécanique analytique* (1788), and the great master was present as far as 1813 to point out its resources and to watch over the legitimacy of its applications. Throughout the whole century each new advance has but vindicated the preëminent power and safety of its methods. It triumphed with Maxwell (1864), when he deduced the concealed kinetics of the electromagnetic field, and with Gibbs (1876–78), when he adapted it to the equilibrium of chemical systems. It will triumph again in the electromagnetic dynamics of the future.

Naturally there were reactions against the tyranny of the method of "liaisons." The most outspoken of these, propounded under the protection of Laplace himself, was the celebrated *Mécanique physique* of Poisson (1828), an accentuation of Boscovich's (1758) dynamics, which permeates the work of Navier, Cauchy, de St. Venant, Boussinesq, even Fresnel, Ampère, and a host of others. Cauchy in particular spent much time to reconcile the molecular method with the Lagrangian abstractions. But Poisson's method, though sustained by such splendid genius, has, nevertheless, on more than one occasion — in capillarity, in elastics — shown itself to be untrustworthy. It was rudely shaken when, with the rise of modern electricity, the influence of the medium was more and more pushed to the front.

Another complete reconstruction of dynamics is due to Thomson and Tait (1867), in their endeavor to gain clearness and uniformity of design, by referring the whole subject logically back to Newton. This great work is the first to make systematic use of the doctrine of the conservation of energy.

Finally, Hertz (1894), imbued with the general trend of contemporaneous thought, made a powerful effort to exclude force and potential energy from dynamics altogether — postulating a universe of concealed motions such as Helmholtz (1884) had treated in his theory of cyclic systems, and Kelvin had conceived in his adynamic gyrostatic ether (1890). In fact, the introduction of concealed systems and of ordered molecular motions by Helmholtz and Boltzmann has proved most potent in justifying the Lagrangian dynamics in its application to the actual motions of nature.

The specific contributions of the first rank which dynamics owes to the last century, engrossed as it was with the applications of the subject, or with its mathematical difficulties, are not numerous. In chronological order we recall naturally the statics (1804) and the rotational dynamics (1834) of Poinsot, all in their geometrical character so surprisingly distinct from the contemporary dynamics of Lagrange and Laplace. We further recall Gauss's principle of least constraint (1829), but little used, though often in its applications superior to the method of displacement; Hamilton's principle of varying action (1834) and his characteristic function (1834, 1835), the former obtainable by an easy transition from D'Alembert's principle and by contrast with Gauss's principle, of such exceptional utility in the development of modern physics; finally the development of the Leibnitzian doctrine of work and *vis viva* into the law of the conservation of energy, which more than any other principle has consciously pervaded the progress of the nineteenth century. Clausius's theorem of the *Virial* (1870) and Jacobi's (1866) contributions should be added among others.

The potential, though contained explicitly in the writings of Lagrange (1777), may well be claimed by the last century. The differential equation underlying the doctrine had already been given by Laplace in 1782, but it was subsequently to be completed by Poisson (1827). Gauss (1813, 1839) contributed his invaluable theorems relative to the surface integrals and force flux, and Stokes (1854) his equally important relation of the line and the surface integral. Legendre (published 1785) and Laplace (1782) were the first to apply spherical harmonics in expansions. The detailed development of volume surface and line potential has enlisted many of the ablest writers, among whom Chasles (1837, 1839, 1842), Helmholtz (1853), C. Neumann (1877, 1880), Lejeune-Dirichlet (1876), Murphy (1833), and others are prominent.

The gradual growth of the doctrine of the potential would have been accelerated, had not science to its own loss overlooked the famous essay of Green (1828), in which many of the important theorems were anticipated, and of which Green's theorem and Green's function are to-day familiar reminders.

Recent dynamists incline to the uses of the methods of modern geometry and to the vector calculus with continually increasing favor. Noteworthy progress was first made in this direction by Moebius (1837–43, *Statik*, 1838), but the power of these methods to be fully appreciated required the invention of the *Ausdehnungslehre*, by Grassmann (1844), and of *quaternions*, by Hamilton (1853). Finally the profound investigations of Sir Robert Ball (1871, *et seq.*, *Treatise*) on the theory of screws with its immediate dynamical applications, though as yet but little cultivated except by the author, must be reckoned among the promising heritages of the twentieth century.

On the experimental side it is possible to refer only to researches of a strikingly original character, like Foucault's pendulum (1851) and Fizeau's gyrostat; or like Boys's (1887, *et seq.*) remarkable quartz-fibre torsion-balance, by which the Newtonian constant of gravitation and the mean density of the earth originally determined by Maskelyne (1775–78) and by Cavendish (1798) were evaluated with a precision probably superior to that of the other recent measurements, the pendulum work of Airy (1856) and Wilsing (1885–87), or the balance methods of Jolly (1881), König, and Richarz (1884). Extensive transcontinental gravitational surveys like that of Mendenhall (1895) have but begun.

Hydrodynamics

The theory of the equilibrium of liquids was well understood prior to the century, even in the case of rotating fluids, thanks to the labors of Maclaurin (1742), Clairaut (1743), and Lagrange (1788). The generalizations of Jacobi (1834) contributed the triaxial ellipsoid of revolution, and the case has been extended to two rotating attracting masses by Poincaré (1885) and Darwin (1887). The astonishing revelations contained in the recent work of Poincaré are particularly noteworthy.

Unlike elastics, theoretical hydrodynamics passed into the nineteenth century in a relatively well-developed state. Both types of the Eulerian equations of motion (1755, 1759) had left the hands of Lagrange (1788) in their present form. In relatively recent times H. Weber (1868) transformed them in a way combining certain advantages of both, and another transformation was undertaken by Clebsch (1859). Hankel (1861) modified the equation of continuity, and Svanberg and Edlund (1847) the surface conditions.

Helmholtz in his epoch-making paper of 1858 divided the subject into those classes of motion (flow in tubes, streams, jets, waves) for which a velocity potential exists and the vortex motions for which it does not exist. This classification was carried even into higher orders of motion by Craig and by Rowland (1881). For cases with a velocity potential, much progress has been made during the century in the treatment of waves, of discontinuous fluid motion, and in the dynamics of solids suspended in frictionless liquids. Kelland (1844), Scott Russel (1844), and Green (1837) dealt with the motion of progressive waves in relatively shallow vessels, Gerster (1804) and Rankine (1863) with progressive waves in deep water, while Stokes (1846, 1847, 1880), after digesting the contemporaneous advances in hydrodynamics, brought his powerful mind to bear on most of the outstanding difficulties. Kelvin introduced the case of ripples (1871), afterwards treated by Rayleigh (1883). The solitary wave of Russel occupied Boussinesq (1872, 1882), Rayleigh (1876), and others; group-waves were treated by Reynolds (1877) and Rayleigh (1879). Finally the theory of stationary waves received extended attention in the writings of de St. Venant (1871), Kirchhoff (1879), and Greenhill (1887). Early experimental guidance was given by the classic researches of C. H. and W. Weber (1825).

The occurrence of discontinuous variation of velocity within the liquid was first fully appreciated by Helmholtz (1868), later by Kirchhoff (1869), Rayleigh (1876), Voigt (1885), and others. It lends itself well to conformal representations.

The motions of solids within a liquid have fascinated many investigators, and it is chiefly in connection with this subject that the method of sources and sinks was developed by English mathematicians, following Kelvin's method (1856) for the flow of heat. The problem of the sphere was solved more or less completely by Poisson (1832), Stokes (1843), Dirichlet (1852); the problem of the ellipsoid by Green (1833), Clebsch (1858), generalized by Kirchhoff (1869). Rankine treated the translatory motion of cylinders and ellipsoids in a way bearing on the resistance of ships. Stokes (1843) and Kirchhoff entertain the question of more than one body. The motion of rings has occupied Kirchhoff (1869), Boltzmann (1871), Kelvin (1871), Bjerknes (1879), and others. The results of C. A. Bjerknes (1868) on the fields of hydrodynamic force surrounding spheres, pulsating or oscillating, in translatory or rotational motion, accentuate the remarkable similarity of these fields with the corresponding cases in electricity and magnetism, and have been edited in a unique monograph (1900) by his son. In a special category belong certain powerful researches with a practical bearing, such as the modern treatment of ballistics by Greenhill and of the ship propeller of Ressel (1826), summarized by Gerlach (1885, 1886).

The numerous contributions of Kelvin (1888, 1889) in particular have thrown new light on the difficult but exceedingly important question of the stability of fluid motion.

The century, moreover, has extended the working theory of the

38   PHYSICS

tides due to Newton (1687) and Laplace (1774), through the labors
of Airy, Kelvin, and Darwin.

Finally the forbidding subject of vortex motion was gradually
approached more and more fully by Lagrange, Cauchy (1815, 1827),
Svanberg (1839), Stokes (1845); but the epoch-making integrations
of the differential equations, together with singularly clear-cut interpretations of the whole subject, are due to Helmholtz (1858). Kelvin
(1867, 1883) soon recognized the importance of Helmholtz's work
and extended it, and further advance came in particular from J. J.
Thomson (1883) and Beltrami (1875). The conditions of stability
in vortex motion were considered by Kelvin (1880), Lamb (1878),
J. J. Thomson, and others, and the cases of one or more columnar
vortices, of cylindrical vortex sheets, of one or more vortex rings,
simple or linked, have all yielded to treatment.

The indestructibility of vortex motion in a frictionless fluid, its
open structure, the occurrence of reciprocal forces, were compared
by Kelvin (1867) with the essential properties of the atom. Others
like Fitzgerald in his cobwebbed ether, and Hicks (1885) in his vortex
sponge, have found in the properties of vortices a clue to the possible structure of the ether. Yet it has not been possible to deduce
the principles of dynamics from the vortex hypothesis, neither is the
property which typifies the mass of an atom clearly discernible.
Kelvin invokes the corpuscular hypothesis of Lesage (1818).

Viscosity

The development of viscous flow is largely on the experimental
side, particularly for solids, where Weber (1835), Kohlrausch (1863,
et seq.), and others have worked out the main laws. Stokes (1845)
deduced the full equations for liquids. Poiseiulle's law (1847), the
motion of small solids in viscous liquids, of vibrating plates, and other
important special cases, has yielded to treatment. The coefficients
of viscosity defined by Poisson (1831), Maxwell (1868), Hagenbach
(1860), O. E. Meyer (1863), are exhaustively investigated for gases
and for liquids. Maxwell (1877) has given the most suggestive and
Boltzmann (1876) the most carefully formulated theory for solids,
but the investigation of absolute data has but begun. The difficulty
of reconciling viscous flow with Lagrange's dynamics seems first to
have been adjusted by Navier.

Aeromechanics

Aerostatics is indissolubly linked with thermodynamics. Aerodynamics has not marked out for itself any very definite line of
progress. Though the resistance of oblique planes has engaged the

116

attention of Rayleigh, it is chiefly on the experimental side that the subject has been enriched, as, for instance, by the labors of Langley (1891) and Lilienthal. Langley (1897) has, indeed, constructed a steam-propelled aeroplane which flew successfully; but man himself has not yet flown.

Moreover, the meteorological applications of aerodynamics contained in the profound researches of Guldberg and Mohn (1877), Ferrel (1877), Oberbeck (1882, 1886), Helmholtz (1888, 1889), and others, as well as in such investigations as Sprung's (1880) on the inertia path, are as yet rather qualitative in their bearing on the actual motions of the atmosphere. The marked progress of meteorology is observational in character.

Acoustics

Early in the century the velocity of sound given in a famous equation of Newton was corrected to agree with observation by Laplace (1816).

The great problems in acoustics are addressed in part to the elastician, in part to the physiologist. In the former case the work of Rayleigh (1877) has described the present stage of development, interpreting and enriching almost every part discussed. In the latter case Helmholtz (1863) has devoted his immense powers to a like purpose and with like success. König has been prominently concerned with the construction of accurate acoustic apparatus.

It is interesting to note that the differential equation representing the vibration of strings was the first to be integrated; that it passed from D'Alembert (1747) successively to Euler (1779), Bernoulli (1753) and Lagrange (1759). With the introduction of Fourier's series (1807) and of spherical harmonics at the very beginning of the century, D'Alembert's and the other corresponding equations in acoustics readily yielded to rigorous analysis. Rayleigh's first six chapters summarize the results for one and for two degrees of freedom.

Flexural vibration in rods, membranes, and plates become prominent in the unique investigations of Chladni (1787, 1796, *Akustik*, 1802). The behavior of vibrating rods has been developed by Euler (1779), Cauchy (1827), Poisson (1833), Strehlke (1833), Lissajous (1833), Seebeck (1849), and is summarized in the seventh and eighth chapters of Rayleigh's book. The transverse vibration of membranes engaged the attention of Poisson (1829). Round membranes were rigorously treated by Kirchhoff (1850) and by Clebsch (1862); elliptic membranes by Mathieu (1868). The problem of vibrating plates presents formidable difficulties resulting not only from the edge conditions, but from the underlying differential equation of the fourth degree due to Sophie Germain (1810) and to Lagrange (1811). The solutions have taxed the powers of Poisson (1812, 1829), Cauchy (1829), Kirchhoff (1850), Boussinesq (1871–79), and others. For the circular plate Kirchhoff gave the complete theory. Rayleigh systematized the results for the quadratic plate, and the general account makes up his ninth and tenth chapters.

Longitudinal vibrations, which are of particular importance in case of the organ-pipe, were considered in succession by Poisson (1817), Hopkins (1838), Quet (1855); but Helmholtz in his famous paper of 1860 gave the first adequate theory of the open organ-pipe, involving viscosity. Further extension was then added by Kirchhoff (1868), and by Rayleigh (1870, et seq.), including particularly powerful analysis of resonance. The subject in its entirety, including the allied treatment of the resonator, completes the second volume of Rayleigh's *Sound*.

On the other hand, the whole subject of tone-quality, of combination and difference tones, of speech, of harmony, in its physical, physiological, and æsthetic relations, has been reconstructed, using all the work of earlier investigators, by Helmholtz (1862), in his masterly *Tonempfindungen*. With rare skill and devotion König contributed a wealth of siren-like experimental appurtenances.

Acousticians have been fertile in devising ingenious methods and apparatus, among which the tuning-fork with resonator of Marloye, the siren of Cagniard de la Tour (1819), the Lissajous curves (1857), the stroboscope of Plateau (1832), the manometric flames of König (1862, 1872), the dust methods of Chladni (1787) and of Kundt (1865–68), Melde's vibrating strings (1860, 1864), the phonograph of Edison and of Bell (1877), are among the more famous.

Heat: Thermometry

The invention of the air thermometer dates back at least to Amontons (1699), but it was not until Rudberg (1837), and more thoroughly Regnault (1841, et seq.) and Magnus (1842), had completed their work on the thermal expansion and compressibility of air, that air thermometry became adequately rigorous. On the theoretical side Clapeyron (1834), Helmholtz (1847), Joule (1848), had in various ways proposed the use of the Carnot function (1894) for temperature measurement, but the subject was finally disposed of by Kelvin (1849, et seq.) in his series of papers on temperature and temperature measurement.

Practical thermometry gained much from the measurement of the expansion of mercury by Dulong and Petit (1818), repeated by Regnault. It also profited by the determination of the viscous behavior of glass, due to Pernet (1876) and others, but more from the elimination of these errors by the invention of the Jena glass.

It is significant to note that the broad question of thermal expansion has yet no adequate equation, though much has been done experimentally for fluids by the magnificent work of Amagat (1869, 1873, *et seq.*).

Heat Conduction

The subject of heat conduction from a theoretical point of view was virtually created by the great memoir of Fourier (1822), which shed its first light here, but subsequently illumined almost the whole of physics. The treatment passed successively through the hands of many of the foremost thinkers, notably of Poisson (1835, 1837), Lamé (1836, 1839, 1843), Kelvin (1841–44), and others. With the latter (1856) the ingenious method of sources and sinks originated. The character of the conduction is now well known for continuous media, isotropic or not, bounded by the more simple geometrical forms, in particular for the sphere under all reasonable initial and surface conditions. Much attention has been given to the heat conduction of the earth, following Fourier, by Kelvin (1862, 1878), King (1893), and others.

Experimentally, Wiedemann and Franz (1853) determined the relative heat conduction of metals and showed that for simple bodies a parallel gradation exists for the cases of heat and of electrical conductivity. Noteworthy absolute methods for measuring heat conduction were devised in particular by Forbes (1842), F. Neumann (1862), Ångström (1861–64), and a lamellar method applying to fluids by H. F. Weber (1880).

Calorimetry

Practical calorimetry was virtually completed by the researches of Black in 1763. A rich harvest of experimental results, therefore, has since accrued to the subjects of specific, latent, and chemical heats, due in particularly important cases to the indefatigable Regnault (1840, 1845, *et seq.*). Dulong and Petit (1819) discovered the remarkable fact of the approximate constancy of the atomic heats of the elements. The apparently exceptional cases were interpreted for carbon silicon and boron by H. F. Weber (1875), and for sulphur by Regnault (1840). F. Neumann (1831) extended the law to compound bodies, and Joule (1844) showed that in many cases specific heat could be treated as additively related to the component specific heats.

Among recent apparatus the invention of Bunsen's ice calorimeter (1870) deserves particular mention.

Thermodynamics

Thermodynamics, as has been stated, in a singularly fruitful way interpreted and broadened the old Leibnitzian principle of *vis viva* of 1686. Beginning with the incidental experiments of Rumford (1798) and of Davy (1799) just antedating the century, the new conception almost leaped into being when J. R. Mayer (1842, 1845) defined and computed the mechanical equivalent of heat, and when Joule (1843, 1845, *et seq.*) made that series of precise and judiciously varied measurements which mark an epoch. Shortly after Helmholtz (1847), transcending the mere bounds of heat, carried the doctrine of the conservation of energy throughout the whole of physics.

Earlier in the century Carnot (1824), stimulated by the growing importance of the steam engine of Watt (1763, *et seq.*), which Fulton (1806) had already applied to transportation by water and which Stephenson (1829) soon after applied to transportation by land, invented the reversible thermodynamic cycle. This cycle or sequence of states of equilibrium of two bodies in mutual action is, perhaps, without a parallel in the prolific fruitfulness of its contributions to modern physics. Its continued use in fifty years of research has but sharpened its logical edge. Carnot deduced the startling doctrine of a temperature criterion for the efficiency of engines. Clapeyron (1834) then gave the geometrical method of representation universally used in thermodynamic discussions to-day, though often made more flexible by new coördinates as suggested by Gibbs (1873).

To bring the ideas of Carnot into harmony with the first law of thermodynamics it is necessary to define the value of a transformation, and this was the great work of Clausius (1850), followed very closely by Kelvin (1851) and more hypothetically by Rankine (1851). The latter's broad treatment of energetics (1855) antedates many recent discussions. As early as 1858 Kirchhoff investigated the solution of solids and of gases thermodynamically, introducing at the same time an original method of treatment.

The second law was not generally accepted without grave misgiving. Clausius, indeed, succeeded in surmounting most of the objections, even those contained in theoretically delicate problems associated with radiation. Nevertheless, the confusion raised by the invocation of Maxwell's " demon " has never quite been calmed; and while Boltzmann (1877, 1878) refers to the second law as a case of probability, Helmholtz (1882) admits that the law is an expression of our inability to deal with the individual atom. Irreversible processes as yet lie quite beyond the pale of thermodynamics. For these the famous inequality of Clausius is the only refuge. The value of an uncompensated transformation is always positive.

The invention of mechanical systems which more or less fully conform to the second law has not been infrequent. Ideas of this nature have been put forward by Boltzmann (1866, 1872), by Clausius (1870, 1871), and more powerfully by Helmholtz (1884) in his theory of cyclic systems, which in a measure suggested the hidden mechanism at the root of Hertz's dynamics. Gibbs's (1902) elementary principles of statistical mechanics seem, however, to contain the nearest approach to a logical justification of the second law — an approach which is more than a dynamical illustration.

The applications of the first and second laws of thermodynamics are ubiquitous. As interesting instances we may mention the conception of an ideal gas and its properties; the departure of physical gases from ideality as shown in Kelvin and Joule's plug experiment (1854, 1862); the corrected temperature scale resulting on the one hand, and the possibility of the modern liquid air refrigerator of Linde and Hampson (1895) on the other. Difficulties encountered in the liquefaction of incoercible gases by Cailletet and Pictet (1877) have vanished even from the hydrogen coercions of Olezewski (1895) and of Dewar and Travers.

Again, the broad treatment of fusion and evaporation, beginning with James Thomson's (1849) computation of the melting point of ice under pressure, Kirchhoff's (1858) treatment of sublimation, the extensive chapter of thermo-elastics set on foot by Kelvin's (1883) equation, are further examples.

To these must be added Andrews's (1869) discovery of the continuity of the liquid and the gaseous states foreshadowed by Cagniard de la Tour (1822, 1823); the deep insight into the laws of physical gases furnished by the experimental prowess of Amagat (1881, 1893, 1896), and the remarkably close approximation amounting almost to a prediction of the facts observed which is given by the great work of van der Waals (1873).

The further development of thermodynamics, remarkable for the breadth, not to say audacity, of its generalizations, was to take place in connection with chemical systems. The analytical power of the conception of a thermodynamic potential was recognized nearly at the same time by many thinkers:[1] by Gibbs (1876), who discovered both the isothermal and the adiabatic potential; by Massieu (1877), independently in his *Fonctions characteristiques ;* by Helmholtz (1882), in his *Freie Energie;* by Duhem (1886) and by Planck (1887, 1891), in their respective thermodynamic potentials. The transformation of Lagrange's doctrine of virtual displacements of infinitely more complicated systems than those originally contemplated, in other words the introduction of a virtual thermodynamic modification in complete analogy with the virtual displacement of the *mécanique analytique,* marked a new possibility of research of

[1] Maxwell's *available energy* is accidentally overlooked in the text.

which Gibbs made the profoundest use. Unaware of this marshaling of powerful mathematical forces, van 't Hoff (1886, 1888) consummated his marvelously simple application of the second law; and from interpretations of the experiments of Pfeffer (1877) and of Raoult (1883, 1887) propounded a new theory of solution, indeed, a basis for chemical physics, in a form at once available for experimental investigation.

The highly generalized treatment of chemical statics by Gibbs bore early fruit in its application to Deville's phenomenon of dissociation (1857), and in succession Gibbs (1878, 1879), Duhem (1886), Planck (1887), have deduced adequate equations, while the latter in case of dilute solutions gave a theoretical basis for Guldberg and Waage's law of mass action (1879). An earlier independent treatment of dissociation is due to Horstmann (1869, 1873).

In comparison with the brilliant advance of chemical statics which followed Gibbs, the progress of chemical dynamics has been less obvious; but the outlines of the subject have, nevertheless, been succinctly drawn in a profound paper by Helmholtz (1886), followed with much skill by Duhem (1894, 1896) and Natanson (1896).

Kinetic Theory of Gases

The kinetic theory of gases at the outset, and as suggested by Herapath (1821), Joule (1851, 1857), Krönig (1856), virtually reaffirmed the classic treatise of Bernoulli (1738). Clausius in 1857–62 gave to the theory a modern aspect in his derivation of Boyle's law in its thermal relations, of molecular velocity and of the ratio of translational to total energy. He also introduced the mean free path (1858). Closely after followed Maxwell (1860), adducing the law for the distribution of velocity among molecules, later critically and elaborately examined by Boltzmann (1868–81). Nevertheless, the difficulties relating to the partition of energy have not yet been surmounted. The subject is still under vigorous discussion, as the papers of Burbury (1899) and others testify.

To Maxwell (1860, 1868) is due the specifically kinetic interpretation of viscosity, of diffusion, of heat conduction, subjects which also engaged the attention of Boltzmann (1872–87). Rigorous data for molecular velocity and mean free path have thus become available, and van der Waals (1873) added a final allowance for the size of the molecules. Less satisfactory has been the exploration of the character of molecular force for which Maxwell, Boltzmann (1872, *et seq.*), Sutherland (1886, 1893), and others have put forward tentative investigations.

The intrinsic equation of fluids discovered and treated in the great paper of van der Waals (1873), though partaking of the character of a first approximation, has greatly promoted the coördination of most of the known facts. Corresponding states, the thermal coefficients, the vapor pressure relation, the minimum of pressure-volume products, and even molecular diameters, are reasonably inferred by van der Waals from very simple premises. Many of the results have been tested by Amagat (1896).

The data for molecular diameter furnished by the kinetic theory as a whole, viz., the original values of Loschmidt (1865), of van der Waals (1873), and others, are of the same order of values as Kelvin's estimates (1883) from capillarity and contact electricity. Many converging lines of evidence show that an approximation to the truth has surely been reached.

Radiation

Our knowledge of the radiation of heat, diathermacy, thermocrosis, was promoted by the perfection which the thermopyle reached in the hands of Melloni (1835–53). These and other researches set at rest forever all questions relating to the identity of heat and light. The subject was, however, destined to attain a much higher order of precision with the invention of Langley's bolometer (1881). The survey of heat spectra, beginning with the laborious attempts of Herschel (1840), of E. Becquerel (1843, 1870), H. Becquerel (1883), and others, has thus culminated in the magnificent development shown in Langley's charts (1883, 1884, *et seq.*).

Kirchhoff's law (1860), to some extent anticipated by Stewart (1857, 1858), pervades the whole subject. The radiation of the black body, tentatively formulated in relation to temperature by Stefan (1879) and more rigorously by Boltzmann (1884), has furnished the savants of the Reichsanstalt with means for the development of a new pyrometry whose upper limit is not in sight.

Among curious inventions Crooke's radiometer (1874) and Bell's photophone may be cited. The adaptation of the former in case of high exhaustion to the actual measurement of Maxwell's (1873) light pressure by Lebedew (1901) and Nichols and Hull (1903) is of quite recent history.

The first estimate of the important constant of solar radiation at the earth was made by Pouillet (1838); but other pyrheliometric methods have since been devised by Langley (1884) and more recently by Ångström (1886, *et seq.*).

Velocity of light

Data for the velocity of light, verified by independent astronomical observations, were well known prior to the century; for Römer had worked as long ago as 1675, and Bradley in 1727. It remained to actually measure this enormous velocity in the laboratory, apparently an extraordinary feat, but accomplished simultaneously by Fizeau (1849) and by the aid of Wheatstone's revolving mirror (1834) by Foucault (1849, 1850, 1862). Since that time precision has been given to this important constant by Cornu (1871, 1873, 1874), Forbes and Young (1882), Michelson (1878, et seq.), and Newcomb (1885). Foucault (1850), and more accurately Michelson (1884), determined the variation of velocity with the medium and wave-length, thus assuring to the undulatory theory its ultimate triumph. Grave concern, however, still exists, inasmuch as Michelson and Morley (1886) by the most refined measurement, and differing from the older observations of Fizeau (1851, 1859), were unable to detect the optical effect of the relative motion of the atmosphere and the luminiferous ether predicted by theory.

Römer's observation may in some degree be considered as an anticipation of the principle first clearly stated by Döppler (1842), which has since become invaluable in spectroscopy. Estimates of the density of the luminiferous ether have been published, in particular by Kelvin (1854).

Geometric optics

Prior to the nineteenth century geometric optics, having been mustered before Huyghens (1690), Newton (1704), Malus (1808), Lagrange (1778, 1803), and others, had naturally attained a high order of development. It was, nevertheless, remodeled by the great paper of Gauss (1841), and was thereafter generalized step by step by Listing, Möbius (1855), and particularly by Abbe (1872), postulating that in character, the cardinal elements are independent of the physical reasons by which one region is imaged in another.

So many able thinkers, like Airy (1827), Maxwell (1856, et seq.), Bessel (1840, 1841), Helmholtz (1856, 1867), Ferraris (1877, 1880), and others have contributed to the furtherance of geometric optics, that definite mention is impossible. In other cases, again, profound methods like those of Hamilton (1828, et seq.), Kummer (1859), do not seem to have borne correspondingly obvious fruit. The fundamental bearing of diffraction on geometric optics was first pointed out by Airy (1838), but developed by Abbe (1873), and after him by Rayleigh (1879). An adequate theory of the rainbow, due to Airy and others, is one of its picturesque accomplishments (1838).

The so-called astronomical refraction of a medium of continuously varying index, successively treated by Bouguer (1739, 1749), Simpson (1743), Bradley (1750, 1762), owes its recent refined development to Bessel (1823, 1826, 1842), Ivory (1822, 1823, et seq.),

Radau (1884), and others. Tait (1883) gave much attention to the allied treatment of mirage.

In relation to instruments the conditions of aplantism were examined by Clausius (1864), by Helmholtz (1874), by Abbe (1873, *et seq.*), by Hockin (1884), and others, and the apochromatic lens was introduced by Abbe (1879). The microscope is still well subserved by either the Huyghens or the Ramsden (1873) eye-piece, but the objective has undergone successive stages of improvement, beginning with Lister's discovery in 1830. Amici (1840) introduced the principle of immersion; Stephenson (1878) and Abbe (1879), homogeneous immersion; and the Abbe-Zeiss apochromatic objective (1886), the outcome of the Jena-glass experiments, marks, perhaps, the high-water mark of the art for the microscope. Steinheil (1865, 1866) introduced the guiding principle for photographic objectives. Alvan Clark carried the difficult technique of telescope lens construction to a degree of astonishing excellence.

Spectrum — Dispersion

Curiously, the acumen of Newton (1666, 1704) stopped short of the ultimate conditions of purity of spectrum. It was left to Wollaston (1802), about one hundred years later, to introduce the slit and observe the dark lines of the solar spectrum. Fraunhofer (1814, 1815, 1823) mapped them out carefully and insisted on their solar origin. Brewster (1833, 1834), who afterwards (1860) published a map of 3000 lines, was the first to lay stress on the occurrence of absorption, believing it to be atmospheric. Forbes (1836) gave even greater definiteness to absorption by referring it to solar origin. Foucault (1849) pointed out the coincidence of the sodium lines with the D group of Fraunhofer, and discovered the reversing effect of sodium vapor. A statement of the parallelism of emission and absorption came from Ångström (1855) and with greater definiteness and ingenious experiments from Stewart (1860). Nevertheless, it was reserved to Kirchhoff and Bunsen (1860, 1861) to give the clear-cut distinctions between the continuous spectra and the characteristically fixed bright-line or dark-line spectra upon which spectrum analysis depends. Kirchhoff's law was announced in 1861, and the same year brought his map of the solar spectrum and a discussion of the chemical composition of the sun. Huggins (1864, *et seq.*), Ångström (1868), Thalén (1875), followed with improved observations on the distribution and wave-length of the solar lines; but the work of these and other observers was suddenly overshadowed by the marvelous possibilities of the Rowland concave grating (1882, *et seq.*). Rowland's maps and tables of the solar spectrum as they appeared in 1887, 1889, *et seq.*, his summary of the elements contained in the sun (1891), each marked a definite stage of advance of the subject. Mitscherlich (1862, 1863) probably was the first to recognize the banded or channeled spectra of compound bodies. Balmer (1885) constructed a valuable equation for recognizing the distribution of single types of lines. Kayser and Runge (1887, *et seq.*) successfully analyzed the structure of the spectra of alkaline and other elements.

The modernized theory of the grating had been given by Rayleigh in 1874 and was extended to the concave grating by Rowland (1892, 1893) and others. A general theory of the resolving power of prismatic systems is also due to Rayleigh (1879, 1880), and another to Thollon (1881).

The work of Rowland for the visible spectrum was ably paralleled by Langley's investigations (1883 *et seq.*) of the infra-red, dating from the invention of the bolometer (1881). Superseding the work of earlier investigators like Fizeau and Foucault (1878) and others, Langley extended the spectrum with detailed accuracy to over eight times its visible length. The solar and the lunar spectrum, the radiations of incandescent and of hot bodies, were all specified absolutely and with precision. With artificial spectra Rubens (1892, 1899) has since gone further, reaching the longest heat-waves known.

A similarly remarkable extension was added for the ultra-violet by Schumann (1890, 1892), contending successfully with the gradually increasing opacity of all known media.

Experimentally the suggestion of the spectroheliograph by Lockyer (1868) and by Janssen (1868) and its brilliant achievement by Hale (1892) promise notable additions to our knowledge of solar activity.

Finally, the refractions of absorbing media have been of great importance in their bearing on theory. The peculiarities of metallic reflection were announced from his earlier experiments (1811) by Arago in 1817 and more fully investigated by Brewster (1815, 1830, 1831). F. Neumann (1832) and MacCullagh (1837) gave sharper statements to these phenomena. Equations were advanced by Cauchy (1836, *et seq.*) for isotropic bodies, and later with greater detail by Rayleigh (1872), Ketteler (1875, *et seq.*), Drude (1887, *et seq.*), and others. Jamin (1847, 1848) devised the first experiments of requisite precision and found them in close agreement with Cauchy's theory. Kundt (1888) more recently investigated the refraction of metallic prisms.

Anomalous dispersion was discovered by Christiansen in 1870, and studied by Kundt (1871, *et seq.*). Sellmeyer's (1872) powerful and flexible theory of dispersion was extended to include absorption effects by Helmholtz (1874), with greater detail by Ketteler (1879, *et seq.*), and from a different point of view by Kelvin (1885).

126

The electromagnetic theory lends itself particularly well to the same phenomena, and Kolázek (1887, 1888), Goldhammer (1892), Helmholtz (1892), Drude (1893), and others instanced its adaptation with success.

Photometry, Fluorescence, Photochemistry

The cosine law of Lambert (1760) has since been interpreted in a way satisfying modern requirements by Fourier (1817, 1824) and by Lommel (1880). Among new resources for the experimentalist the spectrophotometer, the Lummer-Brodhun photometer (1889), and Rood's flicker photometer (1893, 1899), should be mentioned.

Fluorescence, though ingeniously treated by Herschel (1845, 1853) and Brewster (1846, et seq.), was virtually created in its philosophical aspects by Stokes in his great papers (1852, et seq.) on the subject. In recent years Lommel (1877) made noteworthy contributions. Phosphorescence has engaged the attention of E. Becquerel (1859), among others.

The laws of photochemistry are in large measure due to Bunsen and Roscoe (1857, 1862). The practical development of photography from its beginnings with Daguerre (1829, 1838) and Niépce and Fox-Talbot (1839), to its final improvement by Maddox (1871) with the introduction of the dry plate, is familiar to all. Vogel's (1873) discovery of appropriate sensitizers for different colors has added new resources to the already invaluable application of photography to spectroscopy.

Interference

The colors of thin plates treated successively by Boyle (1663), Hooke (1665), and more particularly by Newton (1672, Optiks, 1704), became in the hands of Young (1802) the means of framing an adequate theory of light. Young also discovered the colors of mixed plates and was cognizant of loss of half a wave-length on reflection from the denser medium. Fresnel (1815) gave an independent explanation of Newton's colors in terms of interference, devising for further evidence his double mirrors (1816), his biprism (1819), and eventually the triple mirror (1820). Billet's plates and split lens (1858) belong to the same classical order, as do also Lloyd's (1837) and Haidinger's (1849) interferences. Brewster's (1817) observation of interference in case of thick plates culminated in the hands of Jamin (1856, 1857) in the useful interferometer. The scope of this apparatus was immensely advanced by the famous device of Michelson (1881, 1882), which has now become a fundamental instrument of research. Michelson's determination of the length of the meter in terms of the wave-length of light with astounding accuracy is a mere example of its accomplishments.

Wiener (1890) in his discovery of the stationary light-wave intro-duced an entirely new interference phenomenon. The method was successfully applied to color photography by Lippmann (1891, 1892), showing that the electric and not the magnetic vector is photographically active.

The theory of interferences from a broader point of view, and including the occurrence of multiple reflections, was successively perfected by Poisson (1823), Fresnel (1823), Airy (1831). It has recently been further advanced by Feussner (1880, et seq.), Sohncke and Wangerin (1881, 1883), Rayleigh (1889), and others. The inter-ferences along a caustic were treated by Airy (1836), but the endeavor to reconstruct geometric optics on a diffraction basis has as yet only succeeded in certain important instances, as already mentioned.

Diffraction

Though diffraction dates back to Grimaldi (1665) and was well known to Newton (1704), the first correct though crude interpret-ation of the phenomenon is due to Young (1802, 1804). Independ-ently Fresnel (1815) in his original work devised similar explanations, but later (1818, 1819, 1826) gave a more rational theory in terms of Huyghens's principle, which he was the first adequately to inter-pret. Fresnel showed that all points of a wave-front are concerned in producing diffraction, though the ultimate critical analysis was left to Stokes (1849).

In 1822 Fraunhofer published his remarkable paper, in which, among other inventions, he introduced the grating into science. Zone plates were studied by Cornu (1875) and by Soret (1875). Rowland's concave grating appeared in 1881 ; Michelson's echelon spectrometer in 1899.

The theory of gratings and other diffraction phenomena was exhaustively treated by Schwerd (1837). Babinet established the principle bearing his name in 1837. Subsequent developments were in part concerned with the improvement of Fresnel's method of computation, in part with a more rigorous treatment of the theory of diffraction. Stokes (1850, 1852) gave the first account of the polarization accompanying diffraction, and thereafter Rayleigh (1871) and many others, including Kirchhoff (1882, 1883), profoundly modified the classic treatment. Airy (1834, 1838) and others elabor-ately examined the diffraction due to a point source in view of its important bearing on the efficiency of optical instruments.

A unique development of diffraction is the phenomenon of scat-tering propounded by Rayleigh (1871) in his dynamics of the blue sky. This great theory which Rayleigh has repeatedly improved (1881, et seq.) has since superseded all other relevant explanations.

Polarization

An infinite variety of polarization phenomena grew out of Bartholinus's (1670) discovery. Sound beginnings of a theory were laid by Huyghens (*Traité*, 1690), whose wavelet principle and elementary wave-front have persisted as an invaluable acquisition, to be generalized by Fresnel in 1821.

Fresh foundations in this department of optics were laid by Malus (1810) in his discovery of the cosine law and the further discovery of the polarization of reflected light. Later (1815) Brewster adduced the conditions of maximum polarization for this case.

In 1811 Arago announced the occurrence of interferences in connection with parallel plane-polarized light, phenomena which under the observations of Arago and Fresnel (1816, 1819), Biot (1816), Brewster (1813, 1814, 1818), and others grew immensely in variety, and in the importance of their bearing on the undulatory theory. It is on the basis of these phenomena that Fresnel in 1819 insisted on the transversality of light-waves, offering proof which was subsequently made rigorous by Verdet (1850). Though a tentative explanation was here again given by Young (1814), the first adequate theory of the behavior of thin plates of æolotropic media with polarized light came from Fresnel (1821).

Airy (1833) elucidated a special case of the gorgeously complicated interferences obtained with convergent pencils; Neumann in 1834 gave the general theory. The forbidding equations resulting were geometrically interpreted by Bertin (1861, 1884), and Lommel (1883) and Neumann (1841) added a theory for stressed media, afterwards improved by Pockels (1889).

The peculiarly undulatory character of natural light owes its explanation largely to Stokes (1852), and his views were verified by many physicists, notably by Fizeau (1862) showing interferences for path differences of 50,000 wave-lengths, and by Michelson for much larger path differences.

The occurrence of double refraction in all non-regular crystals was recognized by Haüy (1788) and studied by Brewster (1818). In 1821, largely by a feat of intuition, Fresnel introduced his generalized elementary wave-surface, and the correctness of his explanation has since been substantiated by a host of observers. Stokes (1862, *et seq.*) was unremittingly active in pointing out the theoretical bearing of the results obtained. Hamilton (1832) supplied a remarkable criterion of the truth of Fresnel's theory deductively, in the prediction of both types of conic refraction. The phenomena were detected experimentally by Lloyd (1833).

The domain of natural rotary polarization, discovered by Arago (1811) and enlarged by Biot (1815), has recently been placed in close relation to non-symmetrical chemical structure by LeBel (1874) and van 't Hoff (1875), and a tentative molecular theory was advanced by Sohncke (1876).

Boussinesq (1868) adapted Cauchy's theory (1842) to these phenomena. Independent elastic theories were propounded by MacCullagh (1837), Briot, Sarrau (1868); but there is naturally no difficulty in accounting for rotary polarization by the electromagnetic theory of light, as was shown by Drude (1892).

Among investigational apparatus of great importance the Soleil (1846, 1847) saccharimeter may be mentioned.

Theories

In conclusion, a brief summary may be given of the chief mechanisms proposed to account for the undulations of light. Fresnel suggested the first adequate optical theory in 1821, which, though singularly correct in its bearing on reflection and refraction in the widest sense, was merely tentative in construction. Cauchy (1829) proposed a specifically elastic theory for the motion of relatively long waves of light in continuous media, based on a reasonable hypothesis of molecular force, and deduced therefrom Fresnel's reflection and refraction equations. Green (1838), ignoring molecular forces and proceeding in accordance with his own method in elastics, published a different theory, which did not, however, lead to Fresnel's equations. Kelvin (1888) found the conditions implied in Cauchy's theory compatible with stability if the ether were considered as bound by a rigid medium. The ether implied throughout is to have the same elasticity everywhere, but to vary in density from medium to medium, and vibration to be normal to the plane of polarization.

Neumann (1835), whose work has been reconstructed by Kirchhoff (1876), and MacCullagh (1837), with the counter-hypothesis of an ether of fixed density but varying in elasticity from medium to medium, also deduced Fresnel's equations, obtaining at the same time better surface conditions in the case of æolotropic media. The vibrations are in the plane of polarization.

All the elastic theories essentially predict a longitudinal light-wave. It was not until Kelvin in 1889–90 proposed his remarkable gyrostatic theory of light, in which force and displacement become torque and twist, that these objections to the elastic theory were wholly removed. MacCullagh, without recognizing their bearing, seems actually to have anticipated Kelvin's equation.

With the purpose of accounting for dispersion, Cauchy in 1835 gave greater breadth to his theory by postulating a sphere of action of ether particles commensurate with wave-length, and in this direction he was followed by F. Neumann (1841), Briot (1864), Rayleigh (1871), and others, treating an ether variously loaded with material particles. Among theories beginning with the phenomena observed, that of Boussinesq (1867, *et seq.*) has received the most extensive development.

The difficult surface conditions met with when light passes from one medium to another, including such subjects as ellipticity, total reflection, etc., have been critically discussed, among others, by Neumann (1835) and Rayleigh (1888); but the discrimination between the Fresnel and the Neumann vector was not accomplished without misgiving before the advent of the work of Hertz.

It appears, therefore, that the elastic theories of light, if Kelvin's gyrostatic adynamic ether be admitted, have not been wholly routed. Nevertheless, the great electromagnetic theory of light propounded by Maxwell (1864, *Treatise*, 1873) has been singularly apt not only in explaining all the phenomena reached by the older theories and in predicting entirely novel results, but in harmoniously uniting, as parts of a unique doctrine, both the electric or photographic light vector of Fresnel and Cauchy and the magnetic vector of Neumann and MacCullagh. Its predictions have, moreover, been astonishingly verified by the work of Hertz (1890), and it is to-day acquiring added power in the convection theories of Lorentz (1895) and others.

Electrostatics

Coulomb's (1785) law antedates the century; indeed, it was known to Cavendish (1771, 1781). Problems of electric distribution were not seriously approached, however, until Poisson (1811) solved the case for spheres in contact. Afterwards Clausius (1852), Helmholtz (1868), and Kirchhoff (1877) examined the conditions for discs, the last giving the first rigorous theory of the experimentally important plate-condenser. In 1845–48 the investigation of electric distribution received new incentive as an application of Kelvin's beautiful method of images. Maxwell (*Treatise*, 1873) systematized the treatment of capacity and induction coefficients.

Riess (1837), in a classic series of experiments on the heat produced by electrostatic discharge, virtually deduced the potential energy of a conductor and in a measure anticipated Joule's law (1841). In 1860 appeared Kelvin's great paper on the electromotive force needed to produce a spark. As early as 1855, however, he had shown that the spark discharge is liable to be of the character of a damped vibration and the theory of electric oscillation was subsequently extended by Kirchhoff (1867). The first adequate experimental verification was due to Feddersen (1858, 1861).

The specific inductive capacity of a medium with its fundamental bearing on the character of electric force was discovered by Faraday in 1837. Of the theories propounded to account for this property the most far-reaching is Maxwell's (1865), which culminates in the unique result showing that the refraction index of a medium is the square root of its specific inductive capacity. With regard to Maxwell's theory of the Faraday stress in the ether as compared with the subsequent development of electrostriction in other media by many authors, notably by Boltzmann (1880) and by Kirchhoff (1885), it is observable that the tendency of the former to assign concrete physical properties to the tube of force is growing, particularly in connection with radioactivity. Duhem (1892, 1895) insists, however, on the greater trustworthiness of the thermodynamic potential.

The seemingly trivial subject of pyroelectricity interpreted by Æpinus (1756) and studied by Brewster (1825), has none the less elicited much discussion and curiosity, a vast number of data by Hankel (1839–93) and others, and a succinct explanation by Kelvin (1860, 1878). Similarly piezoelectricity, discovered by the brothers Curie (1880), has been made the subject of a searching investigation by Voigt (1890). Finally Kerr (1875, et seq.) observed the occurrence of double refraction in an electrically polarized medium. Recent researches, among which those of Lemoine (1896) are most accurate, have determined the phase difference corresponding to the Kerr effect under normal conditions, while Voigt (1899) has adduced an adequate theory.

Certain electrostatic inventions have had a marked bearing on the development of electricity. We may mention in particular Kelvin's quadrant electrometer (1867) and Lippmann's capillary electrometer (1873). Moreover, among apparatus originating in Nicholson's duplicator (1788) and Volta's electrophorus, the Töpler-Holtz machine (1865–67), with the recent improvement due to Wimshurst, has replaced all others. Atmospheric electricity, after the memorable experiment of Franklin (1751), made little progress until Kelvin (1860) organized a systematic attack. More recently a revival of interest began with Exner (1886), but more particularly with Linss (1887), who insisted on the fundamental importance of a detailed knowledge of atmospheric conduction. It is in this direction that the recent vigorous treatment of the atmosphere as an ionized medium has progressed, owing chiefly to the indefatigable devotion of Elster and Geitel (1899, et seq.) and of C. T. R. Wilson (1897, et seq.). Qualitatively the main phenomena of atmospheric electricity are now plausibly accounted for; quantitatively there is as yet very little specific information.

Volta Contacts

Volta's epoch-making experiment of 1797 may well be added to the century which made such prolific use of it; indeed, the Voltaic pile (1800–02) and Volta's law of series (1802) come just within it. Among the innumerable relevant experiments Kelvin's dropping electrodes (1859) and his funnel experiment (1867) are among the more interesting, while the *Spannungsreihe* of R. Kohlrausch (1851, 1853) is the first adequate investigation. Nevertheless, the phenomenon has remained without a universally acceptable explanation until the present day, when it is reluctantly yielding to electronic theory, although ingenious suggestions like Helmholtz's *Doppelschicht* (1879), the interpretations of physical chemistry and the discovery of the concentration cell (Helmholtz; Nernst, 1888, 1889; Planck, 1890) have thrown light upon it.

Among the earliest theories of the galvanic cell is Kelvin's (1851, 1860), which, like Helmholtz's, is incomplete. The most satisfactory theory is Nernst's (1889). Gibbs (1878) and Helmholtz (1882) have made searching critical contributions, chiefly in relation to the thermal phenomena.

Volta's invention was made practically efficient in certain famous galvanic cells, among which Daniell's (1836), Grove's (1839), Clarke's (1878), deserve mention, and the purposes of measurement have been subserved by the potentiometers of Poggendorff (1841), Bosscha (1855), Clarke (1873).

Seebeck Contacts

Thermoelectricity, destined to advance many departments of physics, was discovered by Seebeck in 1821. The Peltier effect followed in 1834, subsequently to be interpreted by Icilius (1853). A thermodynamic theory of the phenomena came from Clausius (1853) and with greater elaboration, together with the discovery of the Thomson effect, from Kelvin (1854, 1856), to whom the thermoelectric diagram is due. This was subsequently developed by Tait (1872, *et seq.*) and his pupils. Avenarius (1863), however, first observed the thermoelectric parabola.

The modern platinum-iridium or platinum-rhodium thermoelectric pyrometer dates from about 1885 and has recently been perfected at the Reichsanstalt. Melloni (1835, *et seq.*) made the most efficient use of the thermopile in detecting minute temperature differences.

Electrolysis

Though recognized by Nichols and Carlisle (1800) early in the century, the laws of electrolysis awaited the discovery of Faraday

133

(1834). Again, it was not till 1853 that further marked advances were made by Hittorf's (1853–59) strikingly original researches on the motions of the ions. Later Clausius (1857) suggested an adequate theory of electrolysis, which was subsequently to be specialized in the dissociation hypothesis of Arrhenius (1881, 1884). To the elaborate investigations of F. Kohlrausch (1879, *et seq.*), however, science owes the fundamental law of the independent velocities of migration of the ions.

Polarization discovered by Ritter in 1803 became in the hands of Planté (1859–1879) an invaluable means for the storage of energy, an application which was further improved by Faure (1880).

Steady Flow

The fundamental law of the steady flow of electricity, in spite of its simplicity, proved to be peculiarly elusive. True, Cavendish (1771–81) had definite notions of electrostatic resistance as dependent on length section and potential, but his intuitions were lost to the world. Davy (1820), from his experiments on the resistances of conductors, seems to have arrived at the law of sections, though he obscured it in a misleading statement. Barlow (1825) and Becquerel (1825–26), the latter operating with the ingenious differential galvanometer of his own invention, were not more definite. Surface effects were frequently suspected. Ohm himself, in his first paper (1825), confused resistance with the polarization of his battery, and it was not till the next year (1826) that he discovered the true law, eventually promulgated in his epoch-making *Die galvanische Kette* (1827).

It is well known that Ohm's mathematical deductions were unfortunate, and would have left a gap between electrostatics and voltaic electricity. But after Ohm's law had been further experimentally established by Fechner (1830), the correct theory was given by Kirchhoff (1849) in a way to bridge over the gap specified. Kirchhoff approached the question gradually, considering first the distribution of current in a plane conductor (1845–46), from which he passed to the laws of distribution in branched conductors (1847–48) — laws which now find such universal application. In his great paper, moreover, Kirchhoff gives the general equation for the activity of the circuit and from this Clausius (1852) soon after deduced the Joule effect theoretically. The law, though virtually implied in Riess's results (1837), was experimentally discovered by Joule (1841).

As bearing critically or otherwise on Ohm's law we may mention the researches of Helmholtz (1852), of Maxwell (1876), the solution of difficult problems in regard to terminals or of the resistance of special forms of conductors, by Rayleigh (1871, 1879), Hicks (1883), and others, the discussion of the refraction of lines of flow by Kirchhoff (1845), and many researches on the limits of accuracy of the law.

Finally, in regard to the evolution of the modern galvanometer from its invention by Schweigger (1820), we may enumerate in succession Nobili's astatic system (1834), Poggendorff's (1826) and Gauss's (1833) mirror device, the aperiodic systems, Weber's (1862) and Kelvin's critical study of the best condition for galvanometry, so cleverly applied in the instruments of the latter. Kelvin's siphon recorder (1867), reproduced in the Depretz-D'Arsonval system (1882), has adapted the galvanometer to modern conditions in cities. For absolute measurement Pouillet's tangent galvanometer (1837), treated for absolute measurement by Weber (1840), and Weber's dynamometer (1846) have lost little of their original importance.

Magnetism

Magnetism, definitely founded by Gilbert (1600) and put on a quantitative basis by Coulomb (1785), was first made the subject of recondite theoretical treatment by Poisson (1824–27). The interpretation thus given to the mechanism of two conditionally separable magnetic fluids facilitated discussion and was very generally used in argument, as for instance by Gauss (1833) and others, although Ampère had suggested the permanent molecular current as early as 1820. Weber (1852) introduced the revolvable molecular magnet, a theory which Ewing (1890) afterwards generalized in a way to include magnetic hysteresis. The phenomenon itself was independently discovered by Warburg (1881) and by Ewing (1882), and has since become of special practical importance.

Faraday in 1852 introduced his invaluable conception of lines of magnetic force, a geometric embodiment of Gauss's (1813, 1839) theorem of force flux, and Maxwell (1855, 1862, *et seq.*) thereafter gave the rigorous scientific meaning to this conception which pervades the whole of contemporaneous electromagnetics.

The phenomenon of magnetic induction, treated hypothetically by Poisson (1824–27) and even by Barlow (1820), has since been attacked by many great thinkers, like F. Neumann (1848), Kirchhoff (1854); but the predominating and most highly elaborated theory is due to Kelvin (1849, *et seq.*). This theory is broad enough to be applicable to æolotropic media and to it the greater part of the notation in current use throughout the world is due. A new method of attack of great promise has, however, been introduced by Duhem (1888, 1895, *et seq.*) in his application of the thermodynamic potential to magnetic phenomena.

Magneticians have succeeded in expressing the magnetic distribution induced in certain simple geometrical figures like the sphere, the spherical shell, the ellipsoid, the infinite cylinder, the ring. Green in 1828 gave an original but untrustworthy treatment for the finite cylinder. Lamellar and solenoidal distributions are defined by Kelvin (1850), to whom the similarity theorems (1856) are also due. Kirchhoff's results for the ring were practically utilized in the absolute measurements of Stoletow (1872) and of Rowland (1878).

Diamagnetism, though known since Brugmans (1778), first challenged the permanent interest of science in the researches of Becquerel (1827) and of Faraday (1845). It is naturally included harmoniously in Kelvin's great theory (1847, *et seq.*). Independent explanations of diamagnetism, however, have by no means abandoned the field; one may instance Weber's (1852) ingenious generalization of Ampère's molecular currents (1820) and the broad critical deductions of Duhem (1889) from the thermodynamic potential. For the treatment of æolotropic magnetic media, Kelvin's (1850, 1851) theory seems to be peculiarly applicable. Weber's theory would seem to lend itself well to electronic treatment.

The extremely complicated subject of magnetostriction, originally observed by Matteuci (1847) and by Joule (1849) in different cases, and elaborately studied by Wiedemann (1858, *et seq.*), has been repeatedly attacked by theoretical physicists, among whom Helmholtz (1881), Kirchhoff (1885), Boltzmann (1879), and Duhem (1891) may be mentioned. None of the carefully elaborated theories accoun s in detail for the facts observed.

The relations of magnetism to light have increased in importance since the fundamental discoveries of Faraday (1845) and of Verdet (1854), and they have been specially enriched by the magneto-optic discoveries of Kerr (1876, *et seq.*), of Kundt (1884, *et seq.*), and more recently by the Zeemann effect (1897, *et seq.*). Among the theorie put forth for the latter, the electronic explanation of Lorentz (1898 1899) and that of Voigt (1899) are supplementary or at least not con tradictory. The treatment of the Kerr effect has been systematized by Drude (1892, 1893). The instantaneity of the rotational effect was first shown by Bichat and Blondlot (1882), and this result has since been found useful in chronography. Sheldon demonstrated the possibility of reversing the Faraday effect. Finally terrestrial magnetism was revolutionized and made accessible to absolute measurement by Gauss (1833), and his method served Weber (1840, *et seq.*) and his successors as a model for the definition of absolute units throughout physics. Another equally important contribution from the same great thinker (1840) is the elaborate treatment of the distribution of terrestrial magnetism, the computations of which have been twice modernized, in the last instance by Neumeyer [1] (1880). Magnetometric methods have advanced but little since the time of Gauss (1833), and Weber's (1853) earth inductor remains a standard instrument of research. Observationally, the development of cycles of variation in the earth's constants is looked forward to with eagerness, and will probably bear on an adequate theory of terrestrial magnetism, yet to be framed. Arrhenius (1903) accentuates the importance of the solar cathode torrent in its bearing on the earth's magnetic phenomena.

Electromagnetism

Electromagnetism, considered either in theory or in its applications, is, perhaps, the most conspicuous creation of the nineteenth century. Beginning with Oersted's great discovery of 1820, the quantitative measurements of Biot and Savart (1820) and Laplace's (1821) law followed in quick succession. Ampère (1820) without delay propounded his famous theory of magnetism. For many years the science was conveniently subserved by Ampère's swimmer (1820), though his functions have since advantageously yielded to Fleming's hand rule for moving current elements. The induction produced by ellipsoidal coils or the derivative cases is fully understood. In practice the rule for the magnetic circuit devised by the Hopkinsons (1886) is in general use. It may be regarded as a terse summary of the theories of Euler (1780), Faraday, Maxwell, and particularly Kelvin (1872), who already made explicit use of it. Nevertheless, the clear-cut practical interpretation of the present day had to be gradually worked out by Rowland (1873, 1884), Bosanquet (1883–85), Kapp (1885), and Pisati (1890).

The construction of elementary motors was taken up by Faraday (1821), Ampère (1822), Barlow (1822), and others, and they were treated rather as laboratory curiosities; for it was not until 1857 that Siemens devised his shuttle-wound armature, and the development of the motor thereafter went *pari passu* with the dynamo, to be presently considered. It culminated in a new principle in 1888, when Ferraris, and somewhat later Tesla (1888) and Borel (1888), introduced polyphase transmission and the more practical realization of Arago's rotating magnetic field (1824).

Theoretical electromagnetics, after a period of quiescence, was again enriched by the discovery of the Hall effect (1879, *et seq.*), which at once elicited wide and vigorous discussion, and for which Rowland (1880), Lorentz (1883), Boltzmann (1886), and others put forward theories of continually increasing finish. Nernst and v. Ettingshausen (1886, 1887) afterwards added the thermomagnetic effect.

---

[1] Dr. L. A. Bauer kindly called my attention to the more recent work of A. Schmidt summarized in Dr. Bauer's own admirable paper.

Electrodynamics

The discovery and interpretation of electrodynamic phenomena were the burden of the unique researches of Ampère (1820, *et seq.*, *Memoir*, 1826). Not until 1846, however, were Ampère's results critically tested. This examination came with great originality from Weber using the bifilar dynamometer of his own invention. Grassmann (1845), Maxwell (1873), and others have invented elementary laws differing from Ampère's; but as Stefan (1869) showed that an indefinite number of such laws might be constructed to meet the given integral conditions, the original law is naturally preferred.

Induction

Faraday (1831, 1832) did not put forward the epoch-making discovery of electrokinetic induction in quantitative form, as the great physicist was insufficiently familiar with Ohm's law. Lentz, however, soon supplied the requisite interpretation in a series of papers (1833, 1835) which contain his well-known law both for the mutual inductions of circuits and of magnets and circuits. Lentz clearly announced that the induced quantity is an electromotive force, independent of the diameter and metal and varying, *caeteris paribus*, with the number of spires. The mutual induction of circuits was first carefully studied by Weber (1846), later by Filici (1852), using a zero method, and Faraday's self-induction by Edlund (1849), while Matteuci (1854) attested the independence of induction of the interposed non-magnetic medium. Henry (1842) demonstrated the successive induction of induced currents.

Curiously enough the occurrence of eddy currents in massive conductors moving in the magnetic field was announced from a different point of view by Arago (1824–26) long before Faraday's great discovery. They were but vaguely understood, however, until Foucault (1855) made his investigation. The general problem of the induction to be anticipated in massive conductor is one of great interest, and Helmholtz (1870), Kirchhoff (1891), Maxwell (1873), Hertz (1880), and others have treated it for different geometrical figures.

The rigorous expression of the law of induction was first obtained by F. Neumann (1845, 1847) on the basis of Lentz's law, both for circuits and for magnets. W. Weber (1846) deduced the law of induction from his generalized law of attraction. More acceptably, however, Helmholtz (1847), and shortly after him Kelvin (1848), showed the law of induction to be a necessary consequence of the law of the conservation of energy, of Ohm's and Joule's law. In 1851 Helmholtz treated the induction in branched circuits. Finally

Faraday's "electrotonic state" was mathematically interpreted thirty years later, by Maxwell, and to-day, under the name of electromagnetic momentum, it is being translated into the notation of the electronic theory.

Many physicists, following the fundamental equation of Neumann (1845, 1847), have developed the treatment of mutual and self induction with special reference to experimental measurement.

On the practical side the magneto-inductor may be traced back to d'al Negro (1832) and to Pixii (1832). The tremendous development of induction electric machinery which followed the introduction of Siemens's (1857) armature can only be instanced. In 1867 Siemens, improving upon Wilde (1866), designed electric generators without permanent magnets. Pacinotti (1860) and later Gramme (1871) invented the ring armature, while von Hefner-Alteneck (1872) and others improved the drum armature. Thereafter further progress was rapid.

It took a different direction in connection with the Ferraris (1888) motor by the development of the induction coil of the laboratory (Faraday, 1831; Neef, 1839; Ruhmkoff, 1853) into the transformer (Gaulard and Gibbs, 1882–84) of the arts. Among special apparatus Hughes (1879) contributed the induction balance, and Tesla (1891) the high frequency transformer. The Elihu Thompson effect (1887) has also been variously used.

In 1860 Reiss devised a telephone, in a form, however, not at once capable of practical development. Bell in 1875 invented a different instrument which needed only the microphone (1878) of Hughes and others to introduce it permanently into the arts. Of particular importance in its bearing on telegraphy, long associated with the names of Gauss and Weber (1833) or practically with Morse and Vail (1837), is the theory of conduction with distributed capacity and inductance established by Kelvin (1856) and extended by Kirchhoff (1857). The working success of the Atlantic cable demonstrated the acumen of the guiding physicist.

Electric Oscillation

The subject of electric oscillation announced in a remarkable paper of Henry in 1842 and threshed out in its main features by Kelvin in 1856, followed by Kirchhoff's treatment of the transmission of oscillations along a wire (1857), has become of discriminating importance between Maxwell's theory of the electric field and the other equally profound theories of an earlier date. These crucial experiments contributed by Hertz (1887, *et seq.*) showed that electromagnetic waves move with the velocity of light, and like it are capable of being reflected, refracted, brought to interference, and polarized. A year later Hertz (1888) worked out the distribution of the vectors in the space surrounding the oscillatory source. Lecher (1890) using an ingenious device of parallel wires, Blondlot (1891) with a special oscillator, and with greater accuracy Trowbridge and Duane (1895) and Saunders (1896), further identified the velocity of the electric wave with that of the wave of light. Simultaneously the reasons for the discrepancies in the strikingly original method for the velocity of electricity due to Wheatstone (1834), and the American and other longitude observations (Walker, 1894; Mitchell, 1850; Gould, 1851), became apparent, though the nature of the difficulties had already appeared in the work of Fizeau and Gounelle (1850).

Some doubt was thrown on the details of Hertz's results by Sarasin and de la Rive's phenomenon of multiple resonance (1890), but this was soon explained away as the necessary result of the occurrence of damped oscillations by Poincaré (1891), by Bjerknes (1891), and others. J. J. Thomson (1891) contributed interesting results for electrodeless discharges, and on the value of the dielectric constant for slow oscillations (1889); Boltzmann (1893) examined the interferences due to thin plates; but it is hardly practicable to summarize the voluminous history of the subject. On the practical side, we are to-day witnessing the astoundingly rapid growth of Hertzian wave wireless telegraphy, due to the successive inventions of Branly (1890, 1891), Popoff, Braun (1899), and the engineering prowess of Marconi. In 1901 these efforts were crowned by the incredible feat of Marconi's first message from Poldhu to Cape Breton, placing the Old World within electric earshot of the New.

Maxwell's equations of the electromagnetic field were put forward as early as 1864, but the whole subject is presented in its broadest relations in his famous treatise of 1873. The fundamental feature of Maxwell's work is the recognition of the displacement current, a conception by which Maxwell was able to annex the phenomena of light to electricity. The methods by which Maxwell arrived at his great discoveries are not generally admitted as logically binding. Most physicists prefer to regard them as an invaluable possession as yet unliquidated in logical coin; but of the truth of his equations there is no doubt. Maxwell's theory has been frequently expounded by other great thinkers, by Rayleigh (1881), by Poincaré (1890), by Boltzmann (1890), by Heaviside (1889), by Hertz (1890), by Lorentz, and others. Hertz and Heaviside, in particular, have condensed the equations into the symmetrical form now commonly used. Poynting (1884) contributed his remarkable theorem on the energy path.

Prior to 1870 the famous law of Weber (1846) had gained wide recognition, containing as it did Coulomb's law, Ampère's law,

Laplace's law, Neumann's law of induction, the conditions of electric oscillation and of electric convection. Every phenomenon in electricity was deducible from it compatibly with the doctrine of the conservation of energy. Clausius (1878), moreover, by a logical effort of extraordinary vigor, established a similar law. Moreover, the early confirmation of Maxwell's theory in terms of the dielectric constant and refractive index of the medium was complex and partial. Rowland's (1876, 1889) famous experiment of electric convection, which has recently been repeatedly verified by Pender and Cremieu and others, though deduced from Maxwell's theory, is not incompatible with Weber's view. Again the ratio between the electrostatic and the electromagnetic system of units, repeatedly determined from the early measurement of Maxwell (1868) to the recent elaborate determinations of Abraham (1892) and Margaret Maltby (1897), with an ever closer approach to the velocity of light, was at its inception one of the great original feats of measurement of Weber himself associated with Kohlrausch (1856). The older theories, however, are based on the so-called action at a distance or on the instantaneous transmission of electromagnetic force. Maxwell's equations, while equally universal with the preceding, predicate not merely a finite time of transmission, but transmission at the rate of the velocity of light. The triumph of this prediction in the work of Hertz has left no further room for reasonable discrimination.

As a consequence of the resulting enthusiasm, perhaps, there has been but little reference in recent years to the great investigation of Helmholtz (1870, 1874), which includes Maxwell's equations as a special case; nor to his later deduction (1886, 1893) of Hertz's equations from the principle of least action. Nevertheless, Helmholtz's electromagnetic potential is deduced rigorously from fundamental principles, and contains, as Duhem (1901) showed, the electromagnetic theory of light.

Maxwell's own vortex theory of physical lines of force (1861, 1862) probably suggested his equations. In recent years, however, the efforts to deduce them directly from apparently simpler properties of a continuous medium, as for instance from its ideal elastics, or again from a specialized ether, have not been infrequent. Kelvin (1890), with his quasi-rigid ether, Boltzmann (1893), Sommerfeld (1892), and others have worked efficiently in this direction. On the other hand, J. J. Thomson (1891, et seq.), with remarkable intuition, affirms the concrete physical existence of Faraday tubes of force, and from this hypothesis reaches many of his brilliant predictions on the nature of matter.

As a final commentary on all these divers interpretations, the important dictum of Poincaré should not be forgotten: If, says Poincaré, compatibly with the principle of the conservation of energy and of least action, any single ether mechanism is possible, there must at the same time be an infinity of others.

The Electronic Theory

The splendid triumph of the electronic theory is of quite recent date, although Davy discovered the electric arc in 1821, and although many experiments were made on the conduction of gases by Faraday (1838), Reiss, Gassiot (1858, *et seq.*), and others. The marvelous progress which the subject has made begins with the observations of the properties of the cathode ray by Plücker and Hittorf (1868), brilliantly substantiated and extended later by Crookes (1879). Hertz (1892) and more specifically Lenard (1894) observed the passage of the cathode rays into the atmosphere. Perrin (1895) showed them to be negatively charged. Röntgen (1895) shattered them against a solid obstacle, generating the X-ray. Goldstein (1886) discovered the anodal rays.

Schuster's (1890) original determination of the charge carried by the ion per gram was soon followed by others utilizing both the electrostatic and the magnetic deviation of the cathode torrent, and by Lorentz (1895) using the Zeeman effect. J. J. Thomson (1898) succeeded in measuring the charge per corpuscle and its mass, and the velocities following Thomson (1897) and Wiechert (1899), are known under most varied conditions.

But all this rapid advance, remarkable in itself, became startlingly so when viewed correlatively with the new phenomena of radioactivity, discovered by Becquerel (1896), wonderfully developed by M. and Madame Curie (1898, *et seq.*), by J. J. Thomson and his pupils, particularly by Rutherford (1899, *et seq.*). From the Curies came radium (1898) and the thermal effect of radioactivity (1903), from Thomson much of the philosophical prevision which revealed the lines of simplicity and order in a bewildering chaos of facts, and from Rutherford the brilliant demonstration of atomic disintegration (1903) which has become the immediate trust of the twentieth century. Even if the ultimate significance of such profound researches as Larmor's (1891) *Ether and Matter* cannot yet be discerned, the evidences of the transmutation of matter are assured, and it is with these that the century will immediately have to reckon.

The physical manifestations accompanying the breakdown of atomic structure, astoundingly varied as these prove to be, assume fundamental importance when it appears that the ultimate issue involved is nothing less than a complete reconstruction of dynamics on an electromagnetic basis. It is now confidently affirmed that the mass of the electron is wholly of the nature of electromagnetic inertia, and hence, as Abraham (1902), utilizing Kaufmann's data

(1902) on the increase of electromagnetic mass with the velocity of the corpuscle, has shown, the Lagrangian equations of motion may be recast in an electromagnetic form. This profound question has been approached independently by two lines of argument, one beginning with Heaviside (1889), who seems to have been the first to compute the magnetic energy of the electron, J. J. Thomson (1891, 1893), Morton (1896), Searle (1896), Sutherland (1899); the other with H. A. Lorentz (1895), Wiechert (1898, 1899), Des Coudres (1900), Drude (1900), Poincaré (1900), Kaufmann (1901), Abraham (1902). Not only does this new electronic tendency in physics give an acceptable account of heat, light, the X-ray, etc., but of the Lagrangian function and of Newton's laws.

Thus it appears, even in the present necessarily superficial summary of the progress of physics within one hundred years, that, curiously enough, just as the nineteenth century began with dynamics and closed with electricity, so the twentieth century begins anew with dynamics, to reach a goal the magnitude of which the human mind can only await with awe. If no Lagrange stands toweringly at the threshold of the era now fully begun, superior workmen abound in continually increasing numbers, endowed with insight, adroitness, audacity, and resources, in a way far transcending the early visions of the wonderful century which has just closed.

Arthur Lalanne Kimball
1856–1922

SECTION A—PHYSICS OF MATTER

(*Hall* 11, *September* 23, 10 *a. m.*)

CHAIRMAN: PROFESSOR SAMUEL W. STRATTON, Director of the National Bureau of Standards, Washington.
SPEAKERS: PROFESSOR ARTHUR L. KIMBALL, Amherst College.
PROFESSOR FRANCIS E. NIPHER, Washington University.
SECRETARY: PROFESSOR R. A. MILLIKAN, University of Chicago.

THE RELATIONS OF THE SCIENCE OF PHYSICS OF MATTER TO OTHER BRANCHES OF LEARNING

BY ARTHUR LALANNE KIMBALL

[Arthur Lalanne Kimball, Professor of Physics, Amherst College. b. October 16, 1856, Succasunna Plains, N. J. A.B. Princeton, 1881; Ph.D. Johns Hopkins University, 1884; post-graduate, Johns Hopkins University; Associate Professor of Physics, Johns Hopkins University, 1888–91; Fellow of American Association for the Advancement of Science, and American Physical Society. Author of *Physical Properties of Gases.*]

IT is evident at the outset that it is quite out of the question, in the time at our disposal, to discuss adequately the relation of the physics of matter to the other sciences, even if the speaker were endowed with the requisite omniscience.

For *matter* is the very stuff in which the phenomena of all the natural sciences are manifested, the chemist finds himself confronted at every turn with physical relations which must be taken into account, the astronomer finds his greatest triumph in exhibiting the universe that he explores with the telescope as an harmonious illustration of physical principles, the geologist also hardly faces a single question that does not demand the aid of physics or chemistry in its solution, and even in the biological sciences the laws of matter still condition the phenomena of life.

Perhaps a brief consideration of the interrelations of these sciences may aid us in a clearer perception of their dependence on the physics of matter.

There are *three* sciences that may be said to be especially fundamental, in that they deal with the elements of the universe of phenomena. These are *physics*, which, if we define it somewhat narrowly, deals with all the phenomena that can be exhibited *by* and *through the means of* any one kind of matter, as well as all interactions between different kinds of matter in which each preserves its separate identity; *chemistry*, which has for its province those special phenomena in which one kind of matter is broken up into two or more kinds, or in which the interactions between different kinds of matter result in the formation of a substance different from either of the constituents; and that phase of *biology* which is concerned with the study of the living cell and of the simplest conditions under which matter exhibits the phenomena of life.

It might have been said that *physics* deals with those phenomena exhibited by and through matter when molecular groupings of atoms are not disturbed, while *chemistry* deals with the phenomena of the formation and breaking-up of the molecules. But such a statement is based upon a theory of the structure of matter which in itself calls for explanation, and therefore the previous statement is preferred as being more general and avoiding the theoretical assumptions that are involved in those just given.

If it is asked what constitutes a particular kind of matter, why, for instance, water-vapor is said to be the same substance as water in the liquid form, it may be said that it is because one can be wholly transformed into the other, each is homogeneous, and remains unchanged in its properties during the transforming, and the transformation is unique.

Professor Ostwald has recently given a most interesting statement of the criterion by which a substance or chemical individual may be recognized without the need of any atomic hypothesis. We may summarize his presentation thus: Where two substances are combined as in solution, there will be one and only one proportion between the quantities of the substances for which, on change of state, such as evaporation or crystallization, the vapor or crystals will have the same composition as the remaining substance, while with a greater or less proportion of either ingredient, there will be a change of concentration with change of state. When such a combination retains this property under widely different conditions of temperature and pressure, it is known as a chemical individual or definite compound. If under *no* circumstances it can be broken up into two phases which differ in constitution, it is called an element.

Ostwald remarks, "The possibility of being changed from one phase into another without variation of the properties of the residue and of the new phase is indeed the most characteristic property of a substance or chemical individual, and all our methods of testing the purity of a substance, or of preparing a pure one, can be reduced to this one property."

But returning to our classification, it is seen that physics, chemistry, and biology are the three fundamental natural sciences, each having as its primary object not the mere arrangement and classification of phenomena, but the formation of such a concept of matter in those relations with which it deals, that the varied facts of observation appear as natural and inevitable consequences.

The other sciences are in a certain sense secondary to the three that have been mentioned. Each is concerned with the investigation of some system that is built up out of matter, and involves the same fundamental relations which are the objects of study for the primary sciences, but the secondary science finds its interest not in the materials of which the structure is made, but in the study of the resulting structure itself.

Thus astronomy seeks to describe and make out the past history and future development of the universe of sun and star and planet. The sciences of the earth are concerned with the history of the development of our planet, with the present phenomena of its interior, of its crust, of its surface, and of its atmosphere, while the secondary biological sciences have as their aim to trace the relations of the various forms of life and to follow out the developments of each.

But while each secondary science thus has an aim of its own quite distinct from that of the primary sciences, nevertheless it must be controlled and to some extent guided by the sciences of matter. Thus in almost every science chemical phenomena play a part which must be reckoned with, while physics, dealing as it does with the most universal phenomena of matter, underlies and conditions all the sciences without exception. Therefore it is to be expected that with the development of physics both in discovery and theory there should be a greater or less reaction on the other sciences, for in so far as they depend for their development on the laws of matter they are dependent on the labors of the physicist.

We might therefore expect to find in every science, if we only knew it well enough, a response to every considerable advance in physics. For the advances in a science result not from discovery alone, but from new points of view taken by those who are thinking on its problems; and the ideas of physics, bearing as they may be said to do on the raw material of the other sciences, must in a preëminent degree influence the thinking of workers in all fields.

It deserves to be emphasized that every science is an intellectual structure. Only as this is conceded will science be yielded the lofty and dignified position which is its due. Experiments may be multiplied, facts and data may be accumulated in bewildering numbers, but there is no science without the clear intellectual vision that sees the parts in their dependencies and relations one to another and catches glimpses of the larger unities that run through all.

They are mistaken who think the true scientist less an idealist than is the artist or student of literature, or who think the path of experiment mere drudgery in the accumulation of insignificant facts. The investigator lives in a world of ideas, and in every step of a difficult inquiry he has the buoyant consciousness that he is getting a deeper, truer insight into his science.

This intellectual character of scientific research is well illustrated in the enthusiasm which marked the news of Hertz's discovery of electromagnetic waves. The facts observed might easily have been thought to be in themselves insignificant : a slight spark observed between the ends of a bent wire near a discharging electrified system. There was no thought of a practical application, and yet a wave of almost unprecedented excitement spread among physicists the world over. Nor was it alone admiration for the skill, the insight and grasp of the great experimenter that won the victory, though this had its effect. It was mainly an exultant enthusiasm over the triumph of an idea, the unification of science in the confirmation of Maxwell's great theory.

It is clear, then, that physics may react on the other sciences in a variety of ways, in its *methods* and *appliances*, in its *discoveries*, and in its *ideas and generalizations;* and it is evident, therefore, that we must limit ourselves to a brief consideration of certain phases of the subject. I have, therefore, chosen to present very briefly some considerations relative to theories of matter, for here physics and chemistry come into the closest contact; also to touch upon some other relations of chemistry and geology to physics, that are of particular interest at this present time.

The fundamental problem in the physics of matter is the nature of matter itself. Of course we recognize at the outset the limitations that bound our attempts at a solution. We may hope to reach eventually some conclusion as to the structure of matter, whether homogeneous or molecular or grained, also as to the relative motions of the parts of the molecule and the law of variation of force between them with the distance. But if we seek to go farther and explain the forces acting in and between molecules in terms of what appear to be more simple and general laws, it seems inevitable that a medium must be assumed, the properties of which will depend on what is assumed as a primary postulate. If we accept, as is usually done, the postulate that forces in their last analysis can only be explained when referred to pressures exerted between contiguous portions of some underlying medium, it seems probable that a theory must be adopted something like the vortex atom theory of Lord Kelvin, with its continuous, incompressible, perfectly fluid medium in which vortically moving portions constitute the atoms, or Osborne Reynolds's theory of space as filled with fine hard spherical grains, in which, regions with nonconformity in arrangement, are the atoms of ordinary matter. Though it must be said that the assumed hardness of the ultimate spherules in the latter theory is a property which in itself needs explanation.

Perhaps, however, in laying down the postulate mentioned above we are pushing too far inferences from our superficial experience.

The idea that force must be a pressure between contiguous portions of substance is derived directly from the notion of the impenetrability of matter. This is why the incompressible medium of Lord Kelvin's theory seems so simple a conception; it is the naked embodiment of the idea of impenetrability associated with inertia.

It is entirely natural that such ideas as impenetrability and inertia, borne in upon us as they are by our experience of matter in bulk, should affect our theorizing, but it should never be forgotten that as fundamental postulates they have no more authority than any others that might be assumed that will coördinate the same facts of observation.

But passing from this more speculative region we find a pretty general agreement on the rough outlines of the structure of matter. With one notable exception most physicists and chemists agree in the idea that matter is atomic or molecular in structure, and that these molecules are in a state of more or less energetic translatory motion, bounding and rebounding from each other. This seems to be the mechanical hypothesis which coördinates the largest number of facts.

A portion of matter is conceived as in a condition of equilibrium under three pressures: the cohesive pressure due to mutual attraction between all molecules which are not farther apart than 50 to 100 millionths of a millimeter; the external pressure, which also acts to cause contraction; and the internal pressure, which balances the two former, and is due to a repulsive force called the force of impact, which is usually supposed to be exerted only between contiguous molecules.

In the solid and liquid states the cohesive pressure is usually very great compared with the external pressure. In case of gases it nearly vanishes. The force between molecules is thus conceived as an attraction which increases rapidly as they approach, until at a certain distance it is balanced by a repulsive force which, increasing still more rapidly, is the controlling force at all less distances.

Lord Kelvin has recently followed out a study of equilibrium conditions in a group of atoms which are assumed to have no mutual influence until within a certain distance, then to attract each other with a force that increases as they approach still nearer, rising to a maximum and then diminishing, and finally becoming a repulsion when the atoms are very near. He remarks, "It is wonderful how much toward explaining the crystallography and elasticity of solids, and the thermo-elastic properties of solids, liquids, and gases, we find without assuming in the Boscovitchian law of force more than one transition from attraction to repulsion."

The fundamental soundness of the conception of matter as having a grained structure of some sort seems to be established by the remarkable degree of agreement in the estimates by various physicists of the size of these ultimate particles, meaning by that the smallest distance between their centres as they rebound from each other, especially when it is considered that these results have been reached from so many different points of view, and are based on such a variety of physical data.

As to the structure of the atom itself a most remarkable theory has been recently developed. J. J. Thomson has marshaled the evidence in favor of the theory proposed by Larmor that matter has an electrical basis, and the theory has already been considerably developed by Lorentz and others. There appears to be reason for believing that the corpuscles of the Kathode rays are simply moving charges of negative electricity, their whole apparent mass being due to their relation to the ether, in consequence of which there is a magnetic field around the moving charge having energy dependent on the square of its velocity. The corpuscle, therefore, effectively has mass in consequence of this reaction between it and the ether.

The corpuscles are found always to carry the same charge, whatever the nature of the gas in which the Kathode rays are formed, and whatever the nature of the electrodes — the charge being the same as that given up by the hydrogen atom in electrolysis, while the mass of the corpuscle is about one one-thousandth that of the hydrogen atom.

The energy in the ether associated with the moving corpuscle depends on the size of the corpuscle as well as upon its charge, and it is found that to account for its apparent mass it must be of extremely small size relative to ordinary atomic dimensions.

Professor Thomson suggests that the primordial element of matter is such a negative electron combined with an equal positive charge, the latter being of nearly atomic dimension. An atom of hydrogen may be thought of as made up of nearly a thousand such pairs, the positive charge being distributed throughout a spherical region giving rise to a field of force within it in which the force on a negative corpuscle will be towards the centre and proportional to its distance from the centre. In this field of force the corpuscles are conceived as describing closed orbits with great velocities.

The internal energy of such an atom is conceived as enormous. In case of the atoms contained in a gram of hydrogen Thomson reckons about 10^{19} ergs as the energy received from mutual attractions in the formation of the atoms, an amount of work that would lift a hundred million kilograms, one thousand meters.

The whole mass of the atom is supposed to be due to the *negative* electrons or corpuscles which it contains. As to the *positive* charge, although it determines the apparent *size* of the atom, it appears to make no contribution to its mass.

When such an atom impacts against another, the corpuscles in each will be disturbed by the jar in their orbital motion, and there will be superposed oscillations which will cause radiation of energy.

If a corpuscle escapes from such an atom, the latter will be left with a positive charge, while if an additional free corpuscle is entrapped, the atom will have a negative charge. The conditions of stability of motion of the corpuscles in the atom would thus determine whether in case of electrolysis the substance would appear electro-positive or electro-negative.

J. J. Thomson, Drude, and others have discussed the electric conduction of metals from the standpoint of this theory. Drude states that in non-conductors only bound electrons are present, that is, positive and negative in combination; and that it is these that determine the dielectric constant of the medium and consequently its index of refraction and optical dispersion; while Langevin explains magnetism and diamagnetism.

Thus we have a theory already surprisingly developed which appears to be applicable to explain many of the properties of matter, though it is not clear that it can give an explanation of cohesion and gravitation. A theory of matter, to be accepted as final, must offer some explanation of the relation between the various elements. Many thinkers have been led to look for some primordial element from which the others are derived, influenced on the one hand by the present evolutionary ideas of biology, and on the other by comparison of spectra and by the remarkable tendency towards whole numbers observed in the atomic weights of the elements which Strutt has discussed from the standpoint of the theory of probabilities. Professor Thomson has accordingly shown how atoms of matter containing great numbers of corpuscles may have been evolved from a simpler primordial form containing fewer corpuscles. But though he has made clear how the hydrogen atom with its thousand corpuscles might be the surviving atom having the *least* number of corpuscles, it is not so clear why there might not be atoms having any number of corpuscles greater than that of hydrogen, within certain limits; why none should be found between hydrogen and helium for example. Some kind of natural selection seems to be needed to explain why some atoms having special numbers of corpuscles survive while intermediate ones are eliminated, though probably the answer is to be sought in the conditions of stability of the motions of the corpuscles.

It is an interesting question what would be the effect of change of temperature of the substance on the motions of the corpuscles in this theory. If the corpuscles in the atom were very numerous, all moving in the same orbit at equal distances apart, they would produce almost the effect of a circular current of electricity, — a steady magnetic field and no radiation; and it seems probable that in the actual case the radiation of internal energy is extremely small, and the total internal energy may be supposed to be so enormous compared with the energy of translation of the atom due to temperature that we may expect no appreciable change in the radiation of internal energy of the atom, whatever the temperature may be.

That component of the vibration of a corpuscle which is radial within the atom, and is set up by the impact of one atom against another, seems to furnish the great mass of radiated energy. This radiation must also react on the motion of the atom as a whole, taking away from the translatory energy of the atom.

The question how the Boltzmann law of partition of energy between the various degrees of freedom will apply to molecules made up of such atoms as are here conceived is an interesting and important one. Is it possible that the *cloud*, as Lord Kelvin calls it, resting on the kinetic theory of gases may be dissipated by the new theory?

This theory of the atom seems also to explain the possibility of the production of spectra of great complexity. It is to be hoped that Balmer's formula and Rydberg's laws of the grouping of lines in spectra may be shown to be the natural outcome of the system of vibration possible in such an atom.

We are startled at first by the very audacity of this theory, seeming as it does to upset the old point of view, and seek the explanation of matter and its laws in terms of the properties of ether and electricity, instead of trying to unravel the secrets of electricity and ether in terms of matter and motion.

Only a few years ago it was thought that the electromagnetic theory of light must be rationalized by giving a mechanical explanation of the various phenomena of the ether, or by showing at least that such an explanation was possible. Witness Maxwell's wonderfully ingenious mechanical model illustrating the phenomena of magnetism, induced currents, and the propagation of electromagnetic waves.

But is it necessary to regard the mechanical explanation as the only sound one? If electricity and ether are fundamental entities underlying all matter and material phenomena, is it not more logical to find a basis for the mechanical laws in some more fundamental laws of ether and electricity which must be accepted as the primary postulates?

In all this development of the atomic view of matter, chemistry and physics have gone hand in hand. The atomic theory of Dalton has been the basis on which both sciences have worked. Avogadro's law for gases has been reached not only by chemical evidence, but has been raised to the rank of a mechanical deduction from the kinetic theory. The significance of the arrangement of atoms in the molecule in determining chemical reaction was emphasized and developed by Kekulé, but it was not until 1874 that the space diagrams of molecules of van't Hoff and Le Bel marked a full appreciation of the possibilities of structure in explaining the differences of isomeric forms.

All of these physical and chemical developments of the atomic theory have been in accordance with a general method of scientific procedure which may be called the method of mechanical models. According to this method, an attempt is made to conceive a certain mechanism by which the various phenomena sought to be explained may be imagined to be brought about.

Such a theory of atoms, for example, if perfect, would exhibit all the properties of atoms as direct consequences of the assumed structure. This cannot, however, be taken as proof that the assumption is real, though for the purpose of our thinking such a theory would have all *the value* of reality, since all consequences deduced from it would conform to the facts of observation. And this suggests wherein the great value of such a theory lies, not alone in the large number of observations which it correlates and brings under a few general principles, but in that it suggests the application of experiments and tests of its sufficiency, thereby enlarging and making more precise our knowledge.

Perhaps the most remarkable instance of the application of this method was Maxwell's development of a mechanical model to illustrate the reactions in the electromagnetic field. Working from this model he developed the equations of the field, which later he deduced in a more general way. And Hertz speaking of them says, "We cannot study this wonderful theory without at times feeling as if an independent life and a reason of its own dwelt in these mathematical formulæ; as if they were wiser than we were, wiser even than their discoverer; as if they gave out more than had been put into them."

On which Boltzmann's comment is, "I should like to add to these words of Hertz only this, that Maxwell's formulæ are simple consequences from his mechanical models; and Hertz's enthusiastic praise is due in the first place, not to Maxwell's analysis, but to his acute penetration in the discovery of mechanical analogies." Such an example well illustrates the importance of the method.

But of recent years, the influence of quite a different method has been strongly marked in chemical research. A method in which certain general laws are established and then applied to particular cases by a process of mathematical reasoning, deducing conclusions quite independently of the particular details of the operation by which they are brought about. This method is well illustrated in Professor J. J. Thomson's work on the application of dynamics to problems in physics and chemistry, and in the deductions based on the laws of thermodynamics that have marked the development of the new physical chemistry.

It is under the influence of this method that Professor Ostwald has been led to propose a theory of matter which does not recognize the necessity of any atomic structure whatever. In a recent address, he says, "It is possible to deduce from the principles of chemical dynamics all the stoichiometrical laws; the law of constant proportion, the law of multiple proportion, and the law of combining weights." And he continues, "You all know that up to this time it has only been possible to deduce these laws by the help of the atomic hypothesis. Chemical dynamics has, therefore, made the atomic hypothesis unnecessary for this purpose and has put the theory of the stoichiometrical laws on more secure ground than that furnished by a mere hypothesis." And then farther on he continues, "*What we call matter is only a complex of energies which we find together in the same place.* We are still perfectly free if we like to suppose either that the energy fills the space homogeneously, or in a periodic or grained way; the latter assumption would be a substitute for the atomic hypothesis." And then he adds, "Evidently there exists a great number of facts — and I count the chemical facts among them — which can be completely described by a homogeneous or non-periodic distribution of energy in space. Whether there exist facts which cannot be described without the periodic assumption, I dare not decide for want of knowledge; only I am bound to say that I know of none."

It is interesting and remarkable that this challenge to the atomic theories of matter should come from the side of chemistry, the very science for which the atomic theory of Dalton was conceived. Especially is it remarkable, in view of the measure of success that has attended the explanation of the differences between such forms as right and left rotating tartaric acids on the basis of molecular structure. And it is difficult to see how it is possible to give any satisfactory explanation of these differences, simply on the basis of the laws of energetics applied to a conception of matter as homogeneous.

With reference to the view that "*What we call matter is only a complex of energies which we find together in the same place,*" it may be said that we recognize different forms of energy only in association with matter or ether; as heat, light, chemical energy, kinetical energy, etc. Hence the term, "a complex of energies," can only mean the total energy in a given region, unless we recognize some vehicle, as matter or ether, in which the special manifestations of energy may exist. This seems to be admitted tacitly by Ostwald himself, for a little farther on he says, "The reason why it is possible to isolate a substance from a solution is that the available energy of the substance is at a minimum." He thus distinguishes between the available and the total energy of a portion of matter. But this discrimination can have no meaning unless it is granted that a portion of the energy of a substance is not available. If we ask why it is not available, the answer may be that when a substance passes from one state to another at constant temperature the work that it can do is less than its total intrinsic energy as a consequence of the laws of thermodynamics. The case must therefore be one to which the second law of thermodynamics can apply. That is, it must involve flow of energy by some such process as heat conduction.

It might perhaps be successfully argued that the very existence of such a process implies grained structure of some sort to which a statistical law may apply. However this may be, it is certainly difficult to conceive of energy as existing apart from some vehicle, matter or ether or both as you will; but to conceive of this sublimated energy as in part available and in part non-available is surely quite beyond attainment.

It is with great diffidence that we dissent from the expressed views of one who has done so much for the advance of physical chemistry, and our excuse for entering on the discussion must be that as the latest utterance with regard to matter, and coming from one who has won the right to have his views given a respectful consideration, it seemed more fitting to present this brief and imperfect discussion than to pass them by without comment.

One of the most important reactions of physics upon the other sciences has resulted from the extension of the *thermodynamic* laws to chemical problems which has marked the new physical chemistry, a science which has sprung into being within the last seventeen years and has already, under the leadership of van 't Hoff, Ostwald, Arrhenius, and Nernst, attained a surprising development, and is making itself felt in many other lines of scientific activity, notably in electrochemistry, geology, and biology. The starting-point in this development was the idea conceived by van 't Hoff that Avogadro's law might be so extended as to apply to the case of substances in solution. Just as a gas expands and fills the containing vessel exerting a pressure against its walls, so a salt dissolved in a liquid diffuses uniformly throughout the liquid and exerts a pressure within the liquid tending to expand it. This osmotic pressure, so called, had been measured in certain cases by Pfeffer and de Vries, but it remained for van 't Hoff to show that, as in case of a gas, the pressure was proportional to the absolute temperature and to the number of molecules of the dissolved substance contained in unit volume.

As has so often happened before, the study of the apparent exceptions to the rule led to a second great advance, the theory of electrolytic dissociation proposed by Arrhenius, to account for the observation that in solutions of electrolytes the osmotic pressure was greater than that reckoned on the basis of the number of molecules present, but was to be explained by their dissociation into ions; thus reaching the same conclusion which Clausius had announced in 1857, but affording a method by which the precise amount of the dissociation might be measured. Additional evidence in favor of this theory was afforded by the studies of the electrical conductivity of dilute solutions of electrolytes made by Kohlrausch.

All this was accompanied by an increasing realization of the important relations that might be established by an application of the laws of thermodynamics to chemical problems. Thus van 't Hoff showed in his paper of 1887 that the depression of the freezing-point of a liquid due to a substance in solution depended directly on the osmotic pressure and could be used to measure it; a result which had already been experimentally reached by Raoul.

In this field, Professor J. Willard Gibbs, in whose recent death the world of science has lost a most profound thinker, was a pioneer. His most important contributions to the subject were in two extraordinary papers, *On the Equilibrium of Heterogeneous Substances.* The first of these related to chemical phenomena, while the second was concerned especially with capillarity and electricity.

To quote from a recent writer, "The most essential feature of Gibbs's discoveries consisted in the extension of the notion of thermodynamical potential to mixtures consisting of a number of components, and the establishment of the properties that the potential is a linear function of certain quantities which Gibbs has called the potentials of the components, and that where the same component is present in different phases, which remain in equilibrium with each other, its potential is the same in all the phases, besides which the temperatures and pressures are equal. The importance of these results was not realized for a considerable time. It was difficult for the experimentalist to appreciate a memoir in which the treatment is highly mathematical and theoretical, and in which but little attempt is made to reduce conclusions to the language of the chemist; moreover it is not unnatural to find the pioneer dwelling at considerable length on comparatively infertile regions of the newly explored territory, while fields that were to prove the most productive were dismissed very briefly."

"It was largely due to Professor van der Waals that two new and important fundamental laws were discovered in Gibbs's paper, namely, the phase rule and the law of critical states."

The phase rule has been the guiding principle in some most important studies of chemical equilibrium. It furnishes a clue by which the polymorphism of such substances as sulphur and tin may be scientifically investigated and the conditions of equilibrium between the different polymorphic forms determined. The studies of the case of ferric chloride by Roozeboom, and of the crystallization out of sea-water of the contained salts by van't Hoff and Meyerhoffer indicates the great value of the phase rule in bringing scientific order out of the complicated relations of the various components and phases involved.

Speaking of this department of physical chemistry, van't Hoff remarked, "Since the study of chemical equilibrium has been related to thermodynamics, and so has steadily gained a broader and safer foundation, it has come into the foreground of the chemical system, and seems more and more to belong there." And Ostwald says in answer to the question, "What are the most important achievements of the chemistry of our day? I do not hesitate to answer: chemical dynamics, or the theory of the progress of chemical reaction, and the theory of chemical equilibrium."

These statements, coming from two masters in the field, are most significant of the importance of the introduction of these ideas into chemistry.

The conceptions and methods of physical chemistry have also been most strongly felt in the field of electrochemical theory. To the question what is the nature of electrolysis, Faraday and Hittorf and Clausius had each contributed important elements of the final answer, then came Arrhenius with the theory of electrolytic dissociation, which has proved so fruitful of consequences, not only in the domain of chemistry, but also in biology and in physics.

One of the most interesting scientific questions connected with electrochemistry is the relation between electromotive force and electrolytic separation, and the development of the theory of the voltaic cell. The question of the seat of electromotive force in the cell was for many years the very storm-centre of physical discussion; but from the standpoint of electrolytic dissociation Nernst has supplemented the work of Helmholtz and Gibbs, and out of all has come a theory which, while not perfect, seems to be in its main features on the solid foundation of the conservation of energy and the laws of thermodynamics.

Another important service for which the world of science is indebted to physics is the determination of the absolute zero of temperature in terms of degrees of the ordinary centigrade scale. About a century ago, Dalton, in his new chemical philosophy, adopts $-3000°$ C. as the probable zero of temperature. While Lavoisier and Laplace make various estimates of the zero ranging from 1500 to 3000 degrees below the freezing-point of water. But when the doctrine of energy became firmly established together with the kinetic theory of gases, it was natural that the condition of a gas in which the particles had no energy of motion, and hence no pressure, should have been taken as indicating the absolute zero. But it was Clausius and Lord Kelvin who based firmly on the laws of thermodynamics the absolute scale of temperature, as we know it to-day.

The absolute zero of temperature has to the physicist all the fascination that the North Pole has to Arctic explorers, and is probably even more difficult to attain. Yet steady progress has been made in conquering the difficult territory that lies toward this goal. The experimental efforts to liquefy the more refractory gases showed that far lower temperatures than had previously been reached must be employed; and step by step, following the suggestions of thermodynamics, the means of attaining low temperatures have been improved, at first cooling by adiabatic expansion of more compressible gases, then aided by the sudden expansion of the gas itself which had been compressed and cooled, and then by a continuous self-intensive action, in which the cold produced by the expansion of one portion of the compressed gas was made use of to cool the still unexpanded gas as it approached the point of expansion.

The mere record of the temperatures reached marks a series of triumphs of ingenuity and perseverance. Thus Faraday, in 1845, reached a temperature of − 110 by the use of solid carbon dioxide and ether evaporated at low pressure. Pictet in 1877 reached − 140, and liquefied oxygen under pressure. Olszewski in 1885 obtained a temperature of − 225 by the evaporation of a mass of solid nitrogen. In 1898 Dewar obtained liquid hydrogen boiling at − 252, or only 20.5 above the absolute zero, and later by boiling at reduced pressures he was able to obtain − 259.5 or 13.5 degrees absolute scale, at which point hydrogen is frozen solid.

The attainment of these low temperatures has not alone made possible investigations of the greatest interest to the physicist, such as studies of the magnetic and electric properties of bodies as they approach the absolute zero, but has enabled the effect of extreme cold on chemical actions to be determined, and has led to the interesting conclusion that "The great majority of chemical interactions are entirely suspended." Though it has been shown by Dewar and Moissan that in case of solid hydrogen and liquid fluorine, violent reaction still takes place even at that small remove from the absolute zero.

A very interesting field has also been opened to biological research, in the effect of extreme cold on the vitality of seeds and micro-organisms. It was found, for example, that barley, pea, and mustard seeds steeped for six hours in liquid hydrogen and thus kept at a temperature of minus 252 degrees, showed no loss of vitality. So, also, certain micro-organisms, among others the bacilli of typhoid fever, Asiatic cholera, and diphtheria, were kept by MacFadyen for seven days at the temperature of liquid air without appreciable loss of vitality. It has been suggested by Professor Travers that, "It is

RELATIONS TO OTHER SCIENCES   83

quite possible that if a living organism were cooled only to temperatures at which physical changes, such as crystallization, take place with reasonable velocity, the process would be fatal, whereas, if they were cooled to the temperature of liquid air no such change would take place within finite time, and the organism would survive."

Also the study of the various combinations of carbon and iron that may exist in steel, and the conditions of equilibrium that exist between them has proved a most important investigation in the field of what van 't Hoff calls solid solutions.

Geology, dealing as it does with the greatest variety of physical processes, such as changes of state, fusion, crystallization, solution, conduction of heat, radiation, with complications depending on variations of pressure and temperature, presents many problems for the solution of which the resources of modern physics must be taxed. The fusing-points of the different chief minerals of the earth's crust, the effect of great pressure on their fusing-points and modes of crystallization, the crystallization of the various elementary minerals out of a fused magma also studied at different pressures, the effect of pressure not only on fusing-points, but on the viscosity and rigidity of minerals at high temperature, the heat conductivities of the various substances making the bulk of the earth's crust, all these are questions that must be thoroughly studied to enable the geologist to determine the probable condition both of temperature and pressure which prevailed during the formation of a given rock mass, and to throw light on the great problem of geology, the age of the earth.

To this latter question, physics has already given a tentative answer. Lord Kelvin's discussion, based on the assumption of the earth as a mass cooling from a uniform high temperature, points to a period of between twenty and one hundred million years, within which geologic changes in the crust of the earth must have occurred; while Helmholtz and Kelvin's deduction of the time during which solar radiation can have been of such an intensity that life conditions on the earth were possible gives about twenty million years as the limit.

But later investigations giving new data as to the properties of the materials of the earth's crust, as to the laws of variation of radiation with temperature, and as to absorption and radiation by the solar and earth's atmospheres, will all contribute to modify and make more precise these methods. Already some progress in this direction has been made. A few years ago, Clarence King gave a most interesting and ingenious rediscussion of Kelvin's cooling of the earth method, making use of the determinations made by Barus of the fusing-points of diabase at different pressures, and gives as the most probable result of the method the period of twenty-four million years, a period in close agreement with that found by Helmholtz and Kelvin from the radiation of the sun.

It should be remarked, however, that in discussing the state of things in the earth's interior, where the pressures so far transcend anything that can be approached in the laboratory, such constants as melting-points should be looked on with great suspicion.

Assuming Laplace's law of distribution of density in the earth, the pressure at a depth of one two-hundredth of the earth's radius is 8600 atmospheres, while at the centre of the earth it becomes more than three million atmospheres. Now the largest pressures that have been used in high temperature experiments are less than three thousand atmospheres. It is evident, then, that any conclusion as to melting-points from laboratory data must be violent exterpolations, if deduced for the enormous pressures at depths greater than one one-hundredth of a radius within the earth, where the pressure will be over 17,000 atmospheres.

But not only is there necessarily great uncertainty as to the fusing-points at these great pressures, but it seems probable that such a process as fusion marked by sudden increase in liquidity can hardly take place at all. In the phenomenon of fusion, the equilibrium of a substance may be regarded as conditioned by the external pressure, the cohesive pressure, and the internal pressure due to the translatory kinetic energy of the molecules, which may be called the kinetic pressure. In a state of equilibrium, the external pressure plus the cohesive pressure must equal the kinetic pressure, the last tending to produce expansion, while the two former act to cause contraction. At ordinary atmospheric pressures in the liquid and solid state, the cohesive pressure is enormously greater than the external pressure. In water at ordinary temperatures it is estimated about 6500 atmospheres, while in a solid such as steel it may have a value of perhaps 18,000 atmospheres. And not only is this cohesive force great relatively to the external pressure, but it decreases with great rapidity as the substance expands. Under these conditions it is easy to see that a slight rise in temperature with consequent expansion and weakening of the cohesive pressure while the kinetic pressure is increased may bring the substance to a point of transition, a melting-point or boiling-point where great changes occur within narrow limits of temperature.

But if we conceive the external pressure to be so great that the cohesive pressure is relatively insignificant, then we should not expect to find any sharply marked changes of state for small changes of temperature or pressure.

To make the case definite assume a temperature of 1000 degrees absolute scale, and a pressure of 1,000,000 atmospheres, and suppose the cohesive pressure is 10,000 atmospheres. Under these circumstances a rise in temperature of ten degrees or a one per cent increase in temperature may be expected to produce a one per cent increase in the kinetic pressure at the original volume; but as the external pressure is constant and the cohesion is insignificant, we may expect a one per cent increase in the volume in which the molecular motions take place or an increase in the mean distance between molecules of one third of one per cent. Such an expansion will be accompanied by slightly lessened cohesive force, less rigidity, and less viscosity, probably; but nothing like a sudden change of state is suggested. The fact that at pressures greater than the critical pressures there can be observed no sharp transition from the liquid to the gaseous state with rise of temperature is quite in accord with the above considerations, and it seems probable that in case of solids under great pressure nothing like melting will be observed, but rather a gradual loss of rigidity or transition to great viscosity, and that the viscosity will decrease steadily with rise in temperature.

But a new aspect is now given to the problem of the age of the earth by the discovery of radioactivity and its attendant phenomena. The earth, instead of being thought of as a cooling body, is now conceived as having within itself a source of almost unlimited energy. Locked up in each atom is believed to be a store of energy so vast that the breaking down of comparatively few of them in the radioactive process will supply the known outflow of heat from the earth.

Rutherford has shown that the observed dissemination of radioactive substances in the earth's crust is probably sufficient to account for the outflow of energy from its surface. Thus the method of estimating the age of the earth from the consideration of it as a cooling body, a method which until lately seemed to physicists to be based on essentially sound premises, and deserving of confidence because of its greater simplicity as compared with the methods by which geological and biological estimates are obtained, is now by the very progress of physics itself abandoned as unreliable.

So also has the study of radioactivity thrown new light on the question of the maintenance of the sun's heat. It is now seen that possible atomic transformations accompanied by the liberation of the vast stores of energy locked up within the atoms of matter may permit an enormous extension of the time during which the sun may have been radiating with something like its present intensity.

In conclusion it may be remarked that a new world is opened to the investigator by the discovery of radioactivity. The atoms of matter are no longer thought of as necessarily fixed and unchangeable. Besides the older problems of matter questions now arise as to evidences of atomic disintegration and change from

more complex to less complex forms, and also the possible development of more complex atoms from simpler ones.

Already we begin to see the effect of these recent discoveries
and ideas on other departments of science. The clue at last seems to
have been found to those long-standing enigmas of nature, thunder-
storms, the Aurora Borealis, the zodiacal light, and the tails of
comets.   But these achievements belong perhaps rather to the
realm of the physics of the ether and of the electron, than to that
of the physics of matter.

Francis Eugene Nipher
1847–1926

PRESENT PROBLEMS IN THE PHYSICS OF MATTER

BY FRANCIS EUGENE NIPHER

[Francis Eugene Nipher, Professor of Physics, Washington University, St. Louis, Mo. b. December 10, 1847, Port Byron, N. Y. Phil.B. State University of Iowa, 1870; A.M. State University of Iowa, 1875; LL.D. Washington University, 1905. Instructor in Physics and Chemistry, State University of Iowa, 1870–74; Professor of Physics, Washington University, 1874. Member of Academy of Science of St. Louis, American Physical Society; Fellow of American Society for the Advancement of Science. Author of *Theory of Magnetic Measurements; Introduction to Graphical Algebra; Electricity and Magnetism;* and many scientific papers.]

In dealing with the subject allotted to me by the officers of the Congress, I must say that I have not presumed to solve the problems which present themselves at this time, nor do I feel competent even to state many of them. But it is instructive, in a time like this, to attempt a general survey of some of the great questions of the day, with a view of noting their bearing upon the knowledge of the past. We are continually made to feel that all of our inquiries and results must be reëxamined, and our conclusions broadened and modified by new phenomena.

Charles Babbage, whose last published work was, if I mistake not, a review of the London Exposition of 1851, in the Ninth Bridgewater Treatise, gave incidentally, by way of enforcing his thoughts, a review of his earlier work on calculating-machines. His work covered the simple case of a machine composed of wheels and levers, capable of computing the successive terms of any series. The simplest case is an arithmetical series, the differences between the successive terms being unity. This is the device which we now use in the street-cars for counting fares. He asserted the possibility of making a machine, capable of computing the terms of such a series, or of any other, continuing the operation for thousands of years; and pointed out that the machine may be so designed that it will then compute one single arbitrary term, having no relation to the series which had preceded. It may then resume the former series, or it may begin computing a geometrical series, or a series of squares or cubes of the natural numbers. A scientific investigator, who is not permitted to see the mechanism, begins to observe and record the series of numbers which are being disclosed on the dials. He soon learns the mathematical law of the series. He observes the time-sequence of the successive terms, and computes the date when this order of things began. He then makes use of his knowledge of other machinery, and makes a working drawing of the hidden mechanism which produces these results. He verifies his work by years of subsequent observations. With what amazement does he finally behold that single arbitrary term! With what amazement does he then see the machine begin to compute the squares or the cubes of the numbers it had previously disclosed! The date when that machine was created and set to work has been rudely called in question by the new and seemingly lawless behavior of which it appears to be capable. And yet the observer still feels that the principles of mechanism have not been shaken by this unlooked-for disclosure. He again begins his work, with broader conceptions of the plan of this machine. And his subsequent work is along precisely the same lines, and by the same methods as his previous work.

It is in exactly this way that all scientific work has proceeded, and I wish to point out a few interesting cases of this kind. I find it impossible to do this without presenting the present aspect of these problems in connection with the work of the past. This plan gives a perspective which not only adds to the interest but to the clearness of the presentation.

The nebular hypothesis was an attempt by Kant, Laplace, and Herschel to trace the evolution of the solar system from a glowing mass of incandescent vapor or gas. As the theory was considered and developed, an immense number of correlated phenomena were found to be in harmony with this hypothesis, and a few discordant phenomena were also found. The operation was, moreover, based on a few fundamental and well-established laws, governing the present condition of the system; such as gravitation, radiation of heat, etc. The case became more and more convincing, as the knowledge of the last century was applied. All of this caused the astronomers and physicists to find it very easy to give to the hypothesis their tacit assent.

Later, Sir William Thomson, now Lord Kelvin, took up the question of underground temperature, and determined the limit in time since which the earth must have begun to solidify. He also assumed that the present order of things had come down to us from the past, and that the present order of things consisted in the radiation of heat from a cooling earth.

The time-interval which Kelvin thus determined was in entire harmony with the nebular hypothesis, but the results were received with something like consternation by geologists, and those who had followed Darwin in the study of the evolution of organic life upon the earth. Afterwards Kelvin sought to show that the process of solidification might have required but a short interval of time, and the evolutionists have found that evolution goes on by steps or sudden changes rather than by a continuous succession of imperceptible increments.

The geologists have never been reconciled to Kelvin's results, and their protests have of late seemed to be on the increase. Of late the situation has changed in various ways. The discovery of radioactive matter in wide diffusion in the earth's crust has reopened the whole question of underground temperature as related to the age of the earth and its past history. Nevertheless, if the nebular theory in any form, or any similar theory, represents the process of evolution of the solar system, a large amount of heat due to gravitational contraction must have resulted, and must have been disposed of by radiation.

During several years I have been giving attention to the conditions of evolution of a gaseous nebula. The equations of equilibrium for such a mass have been developed.[1] A cosmical mass of gas was assumed, satisfying everywhere the Boyle-Gay-Lussac law, capable therefore of expanding, of being compressed, and of transmitting pressure, and having a centre towards which it gravitates.

Such a mass of gas is a simple heat-engine. The piston face is any spherical concentric surface. The load on the piston is the weight of superposed layers, external to the piston face. The radially inwardly directed pressure is exactly that required to balance the outward pressure of the inclosed mass. As radiation and contraction proceed, the load on the piston increases, in a perfectly definite way, due to increase in weight of each element of mass as it approaches the gravitating centre. Whatever may be the nature of the gas, as determined by the numerical value of the Boyle-Gay-Lussac constant, at some time in its history contraction will have proceeded until some fixed or definite mass shall have been compressed within a fixed volume of definite radius. The equations show that the pressure at the surface of this mass, that is to say, the load on the piston, will then be entirely independent of the nature of the gas.

The difference between gases will only be shown in the time required for them to reach this assumed stage in their gravitational history. A gas which permits the heat of compression within the piston face to escape most quickly into the refrigerator external to the nebula will reach this stage most quickly. When this has been done, pressures and densities at the piston face are wholly independent of the nature of the gas. The total work of compression done on the mass within the piston face up to this time is also independent of the nature of the gas. But the temperatures at the piston face will be inversely as the numerical value of the Boyle-Gay-Lussac constant.

It is evident, therefore, that the law of contraction cannot be indeterminate as in the case where the load is imposed by the hand of man. There is, therefore, in addition to the Boyle-Gay-Lussac law, another definite relation between any two of the three variables involved in that law. The application of well-known equations of

[1] *Transactions*, Academy of Science of St. Louis, XIII, no. 3; XIV, no. 4.

thermodynamics led to the result that the density at any such piston face was directly proportional to the nth power of the pressure. The value of n is found to be 0.908 for all gases like oxygen, hydrogen, nitrogen, and air. The operation is, therefore, one lying between isothermal and isentropic compression, and near to the former. The specific heat of gravitational compression is therefore negative. The unit mass of gas at any point rises in temperature during compression, and for a rise of temperature of 1°C., it gives off by radiation a definite amount of heat.

If, now, such a nebula be supposed to extend to an infinite distance from the gravitating centre, the mass of the nebula will be infinite. Pressure, density, and temperature then all become zero at an infinite distance. Suppose such a nebula to have reached such a stage in its contraction that the mass of our solar system, 1.99×10^{33} grammes, is internal to Neptune's orbit, then it turns out that the pressure there will be about what it is in Crookes tube, 1.74×10^{-7} atmospheres. The density will be far less than in a Crookes tube, viz.: 1.40×10^{-12} c. g. s. The temperature for a hydrogen nebula will be 3000°C., and for other gases it will be higher in inverse ratio as the value of the Boyle-Gay-Lussac constant.

If the mass of the nebula be made finite, the conditions become still more interesting. Let the condition be imposed that the mass of the nebula is that of our solar system, and that it has so contracted that Neptune's mass only is external to Neptune's orbit. Then the temperature at Neptune's place drops to about 1900°C., for hydrogen,[1] and both pressure and temperature become very much less than before. P 1.49×10^{-10}; d 1.93×10^{-15}. The thickness of the spherical shell which would contain Neptune's mass is about a million miles (1.65×10^{-11} cm.). At the external surface of this nebula, the condition imposed makes P, d, and T zero, as the equations show. Nevertheless, a large fraction of Neptune's mass would be gaseous and far above its critical temperature. It seems to me impossible to think of a nebula having such properties generating by any reasonable rotation a system of planetary bodies. With Neptune's mass on the surface of such a nebula consisting of matter having a density and pressure less than a thousandth of these values in a Crookes tube vacuum, how could we conceive of this matter being gathered into a single planet?

A much more reasonable hypothesis is one discussed by G. H. Darwin in 1889, in the Philosophical Transactions of the Royal Society.[2] Darwin discussed the properties of a swarm of solid meteoric masses, and gives very strong proof of the proposition that

[1] In a nebula of mixed gases, each gas will, of course, have its own temperature, as is well understood.
[2] On the "Mechanical Conditions of a Swarm of Meteorites," and on "Theories of Cosmogony," *Phil. Trans.* 1889.

a system of planetary bodies may originate in this way, although he is very cautious and conservative in stating conclusions. The great importance of this theory of planetary origin from the standpoint of planetary geology and the evolution theory seems to demand that it should receive more attention than it has yet received. The temperature of the great mass of such a swarm will be very much lower than in the case of the gaseous nebula. The larger part of such a mass will approach absolute zero in temperature. According to this hypothesis, even Mercury may have been solid when it separated from the parent mass, although in its later stages a large mass might become a gaseous nebula, as the sun now is. But in case of a body like our earth, of such relatively small size, and so far removed from the heated core, there does not seem to be any necessity for the assumption that it was ever in a fused condition.

In view of these new developments, it seems peculiarly important that a discussion of the limits of maximum temperature which the mass of our earth has reached in the past should now be taken in hand again. Suppose a swarm of meteorites to fill the space internal to the moon's orbit, having a total mass equal to that of our earth. Assume that the mass is in rotation, so that the moon is about to separate from the parent mass. It would probably be too radical to assume that each element of mass has either the same actual velocity or the same angular velocity. Various hypotheses, more or less probable, are possible. Assume an initial temperature approaching zero absolute. It seems clear that the highest temperature reached in passing to the present condition of things may be far below the temperature of fusion.

A body falling directly from the moon's distance to the earth will develop 59/60 of the kinetic energy it would acquire in falling from an infinite distance. The earth is yet being bombarded by meteoric matter having such velocities. But the operation is taking place so slowly that the heat has time to become dissipated by radiation, so that no appreciable rise in temperature of the earth results. To what extent may this condition have held in the past? Darwin discussed the tendency of the larger masses in such a swarm to accumulate towards the centre. It is a kind of sorting process. These larger masses would be in general of a metallic character. The more brittle rocks of smaller density would therefore form the outer layers of our earth. May not the heterogeneous character of our so-called igneous rocks be explained in this way? And the shrinking of the earth would then perhaps be in part the flowing of this porous mass into continuity. And it may incidentally be pointed out that the existence of the belt of meteorites known as the asteroids is a most significant indication of the conditions which must have existed at a certain stage in the history of our solar system.

The problems of the present which have aroused general interest are those which pertain to the physical constitution of matter. And here we are at once confronted with the question, What do we mean by matter? How is matter to be recognized? Of late we have been hearing such phrases as "the electrical theory of matter." There seems to be a marked tendency towards the idea that matter and its properties are alike electrical phenomena. Some even intimate that the molecular theory of gases, and the atomic theory of the chemist are tottering to a fall. We have long known that matter in motion is a form of energy. This energy of moving matter is continually being converted into molecular or atomic vibration, and then escapes from us, apparently, forever, in the form of ether waves. We have also long known that electricity in motion is a form of energy, and that the energy so manifesting itself is also all finally converted into heat, and then into ether waves.

Now this parallel certainty suggests an electrical theory of matter, but it also suggests, equally, a material theory of electricity. And so far from being antagonistic, these two theories are identical. There is nothing whatever to show that electricity has ever been separated from something which has what we have been accustomed to call mass. Rowland [1] found that when the charged sectors on his rotating disk were rotated, a magnetic field was produced, corresponding to that produced by a current of electricity. If the motion of the matter which carries the positive electric charge is in a positive direction, the field is the same as that produced when a negative charge is moved in a negative direction.

Rutherford has recently found phenomena of radioactive matter which have a most vital interest in connection with Rowland's work. The a and β particles which are shot off from such matter are moving in the same direction, and they are oppositely deflected in a magnetic field. They behave like superposed or perhaps juxtaposed electric currents of opposite sign flowing in the same direction. If in these radiations the a and β particles were moving in opposite directions, then in a magnetic field they would be deflected in the same direction. This at once raises a question concerning the nature of an electric current in a conducting wire. Let us assume that we start with the positive and negative charges on the terminals of the Holtz machine. What is it that is taking place when the terminals are joined by wires leading to a galvanometer? We get a current which we are wont to say is due either to a positive current flowing in a positive direction, or to a negative current flowing in an opposite direction. If we cease to apply work to the rotating wheel, it comes to rest, and the potential of the conducting wire becomes uniform throughout. Its extremities which terminate in front of the charged

[1] *American Journal of Science*, [3] **xv**, 30–38, 1878.

inductors are therefore so charged as to produce this uniform potential in the presence of these charged inductors, and the polarized glass of the rotor. The ends of the conductor are therefore oppositely charged. There is on its surface a neutral line of no charge. During the motion of the rotor these opposite charges are oppositely directed in the conductor. They are continually being added together. Equal quantities of unlike signs are continually being added together. Are we to assume that equal currents of unlike signs are superposed? Is a positive current in a positive direction identical with a negative current in a negative direction? Mathematically we should say yes. The resulting current, moreover, is uniform throughout the circuit, when measured by its external electromagnetic effects. We may loop in calibrated galvanometers at any point in the circuit, and they tell the same story. But what do the results of Rowland and Rutherford teach us? The β particles carry the negative charge. The negative charge is part and parcel of something which has a positive mass. The a particles are perhaps a combination of more β particles in combination with other particles having (or being) a positive charge of greater numerical value. We have found long ago that the products of an explosion are not necessarily composed of matter in its most elementary form. But these a particles are also part and parcel of something which has a positive mass.

Are we to think of this conductor as being the seat of some action by which positive masses are being urged in a positive direction and positive masses are also being urged in an opposite direction? Are we to think that the mass of such a conductor, carrying a direct current, is slowly increasing, and that after many thousands of years this increase will become appreciable, resulting, perhaps, in a clogging of the conductor, and a decrease in its conduction? In that case a current of positive electricity moving in a positive direction is not a current of negative electricity moving in a negative direction. In that case the nature of positive and negative currents of electricity flowing in opposite directions is fundamentally different from that of the flow of heat and cold in opposite directions, for it involves the motion of masses in opposite directions. It would be interesting to examine whether the long-continued use of a conductor carrying a continuous current may not result in conferring upon it radioactive properties. The results of J. J. Thomson [1] on the phenomena shown by a Geissler tube 15 meters in length are very significant in this connection. He finds the positive luminescence to travel in a direction opposite to that of the cathode stream in the Crookes tube, with a velocity somewhat more than half that of light. The older results of Wheatstone [2] also show that the current from a Leyden jar travels in

[1] *Recent Researches in Electricity and Magnetism*, p. 116.
[2] *Phil. Trans.*, Royal Society, London, 1834.

opposite directions within the conductor which joins its coatings. The middle point of the conductor is last reached by the discharge. If the discharge is maintained and a steady current is finally produced, this current must apparently consist of positive and negative electricity flowing in opposite directions.

If air be pumped out of one boiler and into another, two kinds of pressure are thus generated. If these pressures are added together, by connecting the boilers by means of a conductor, these pressures are added together, and both disappear. If we tap these charged boilers, the discharge from one will attract, and from the other will repel, an uncharged testing sphere. If the testing sphere be itself charged, we shall find that like charges repel, if both are positive, and attract, if both are negative.

It is unnecessary here to enlarge upon the well-known differences between the positive and negative terminals of an exhausted tube. All of these phenomena will finally be helpful in arriving at the nature of the difference between positive and negative electricity. But I will refer to certain phenomena which do not seem to be so well known. Every one is familiar with the small points of light which may often be seen dancing in a crazy fashion over the cathode knob of the Holtz machine. A similar appearance can be seen on the negative carbon of a direct current arc, and in the negative bulb of the mercury vapor-lamp. These points of light may be made to pass from the cathode knob of the Holtz machine to the surface of a photographic dry-plate, exposed in open daylight.[1] Separate the knobs so that no spark will pass. Place the plate near or between them. Connect the knobs with two small metal disks, each armed with a pin-point, so bent that it makes contact with the film. The point of the pin may rest upon the short mark of a lead pencil, drawn upon the film, the pins pointing towards each other on the plate. Points of light, like the so-called ball-lightning discharges, will come from the cathode terminal and successively travel slowly over the plate, leaving a blackened trail of reduced silver behind. By means of a lead pencil held in the hand with the point near the cathode pin-point, these discharges may be induced to make their appearance on the film, and may be deflected into various directions after they have appeared. When left to themselves these minute specimens, of what may perhaps be called ball-lightning, tend to follow the lines of the field, but their paths are somewhat affected by the paths of prior discharges. If one of these points of light is seen on the pin which arms the cathode terminal, there will usually be none upon the film of the dry-plate. It may be brought upon the plate by holding a pencil-point near it.

These ball discharges come from the cathode and travel to or towards the anode. They cannot be induced to come from the anode,

---

[1] *Transactions* of the Academy of Science of St. Louis, x, no. 6.

or to travel against the negative current. The anode terminal has a visible discharge which appears to pass from it, and the photographic plate at the anode looks somewhat like a picture of a relief map of the delta formation at the mouth of a river.

If a conductor be laid upon the plate between the two pin-points, there are then two gaps in the circuit. Each has an anode and a cathode. This conductor may be a metal disk armed with pins 180 degrees apart, which face the discharge points. It may be a pencil-mark upon the film or even a spot of reduced silver on the film. The same discharge will start from the cathode terminal of this intermediate conductor and will travel slowly in the negative direction.

With an induction coil giving an eight-inch spark, these ball discharges can be formed on the surface of wood. In all cases it is evident that chemical work is being done by the slowly advancing ball or point of light, and it is interesting to observe that it is the cathode discharge only which seems to be active. The reason for this may be partly electrical and partly chemical. The anode terminal of the machine may be grounded on a gas-pipe, and the cathode terminal only armed with a point, and the plate may be placed far away from the machine, connection being made between its cathode terminal and the pin-point on the film, with the same results. It may be added that these plates may be of the most sensitive character, and may be freely exposed to daylight for days before they are used. They may also be developed in the light in a bath not very strongly alkaline. The plate will develop clear, with the discharge tracks dark. The picture will not reverse photographically. It probably would do so if the plate were exposed to direct sunlight while the electrical exposure is made.

With an induction coil having an alternating potential on its terminals, these ball discharges may be obtained from both terminals. They will travel towards each other if on the same plate, but they will not unite.

In a closed circuit, one part of which is moved across the lines of a magnetic field, as in the case of a dynamo, we must suppose that the positive and negative currents, if both exist, are superposed in that part of the wire in which the electromotive force originates. The currents are superposed at their origin. The same ether machinery which urges the positive current in one direction urges the negative current in the opposite direction. With the Holtz machine, we have one half of the machine positively and the other half negatively charged. If the knobs are widely separated, and conductors each armed with the pin-point be led off in opposite directions, each terminating on the film of a photographic plate, the cathode will deliver a ball discharge upon its film, while the anode will not. The machine terminal which is not being used may, if desired, be grounded on a gas-pipe. If a pencil-mark be made upon the plate near the anode it will be acted upon inductively, and a ball discharge will pass from it to the anode pin-point. The positive discharge will go in the opposite direction from the pencil-mark, but it leaves no trace. It appears that this ball discharge upon the surface, which results in a destruction of the insulation of the surface, is a characteristic of the negative current.

What would be the result if a suspended Maxwell coil were to be looped into either of these unipolar circuits? Would this case necessarily give the same result that Maxwell obtained?[1] Of course we know that the result which Maxwell sought to detect is very small. We are more particularly concerned with the nature of the action than with the magnitude of the result. If the a particles are so large that they can contribute little or nothing to the current through a metallic conductor, then the positive current may practically be left out of consideration. But it seems doubtful whether the a particles are ultimate in their character, and here is where experimental work is yet needed. It would be exceedingly interesting to study these ball discharges upon a photographic plate under diminishing pressures, as they gradually become a cathode discharge, in a Crookes tube. A Crookes tube may be connected by only one of its terminals to the Holtz machine. The free terminals of the machine and tube may be connected to wires hung on silk fibres and making contact with many pointed ground plates hung on long silk fibres in air. The terminals are then in fact grounded on the dust particles in the air. Either one of these air contacts may be replaced by a ground on the gas-pipe. In all of the possible arrangements covered in this description the tube will give excellent X-ray pictures.

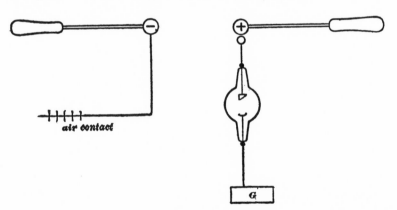

One of these arrangements is represented in the annexed figure, where the cathode terminal of the tube is grounded on the gas-pipe,

[1] *Electricity and Magnetism*, II, p. 201.

and is therefore at zero potential. The ground contact of the tube may be replaced by an air contact, and the negative terminal of the machine may then be grounded on the gas-pipe if desired. In none of these cases are the positive and negative currents delivered by the machine superposed in the X-ray tube. In all these cases X-ray effects are obtained, but in some respects the tube behaves very differently when in the positive current from what it does in the negative. In the negative unipolar circuit, the cathode terminal of the tube is in direct communication with the negative terminal of the machine. When in the positive circuit, the anti-cathode terminal of the tube is in direct communication with the positive terminal of the machine, and the cathode terminal is acted upon inductively across the Crookes tube vacuum. In the negative current the luminous appearances are normal and stable. When in the positive current, the discharge may be made to cease by holding the hands near the bulb, and the luminous glow is affected by the motion of neighboring bodies. The discharge is much more unstable. If the observer approaches the suspended grounding device the face and the hands are covered by luminous points of light, which characterize the cathode terminal. This phenomenon is very striking under these conditions. Ball discharges may be drawn from such point discharges on a metal point to a photographic plate, moving on the plate towards the anode wire or contact plate suspended in air. It is apparent that the wire, when considerably removed from other bodies, is discharging upon the dust particles in the air.

In 1879 Spottiswoode and Moulton[1] published a paper containing a great array of experiments upon the spark discharge through gases. They there dealt with unipolar discharge, and their conclusions are well worthy of notice in this connection. They conclude that "the independence of the discharge from each terminal of the tube is so complete that we can at will cause the discharges from the two terminals to be equal in intensity but opposite in sign (as in the case of the coil) or of any required degree of inequality (as in the case of the coil with a small condenser). Or we can cause the discharge to be from one terminal only, the other terminal acting merely receptively (as in the case of the air-spark discharge with the Holtz machine); or we can cause the discharge to pass from one terminal only, and return to it, the other terminal not taking any part in the discharge; or finally, we can make the two terminals pour forth independent discharges of the same sign, each of which passes back through the terminal from whence it came." This work was done before the Crookes tube had appeared. It is certainly interesting to observe that when a high degree of rarefaction has been reached, the activity within the tube is represented by the cathode stream,

[1] *Phil. Trans.*, 1879.

even when the terminal from which it comes is acted upon only inductively. The remarkable thing is that the X-ray effects and the luminosity of the tube should then be so great. The unipolar positive discharge in the positive direction and the unipolar negative discharge in the negative direction give, in the same time, an X-ray picture of the same intensity, when developed in the same bath, although in the latter case the cathode is only acted on inductively.

Unquestionably the great problem of to-day is the determination of the nature of positive electricity and its relation to what is left when the β particles have been removed. When the cathode particles have left the induced cathode terminal it is positively charged, and communicates that charge to the dust in the air, or to neighboring bodies. It does this, however, by a similar inductive action, and the ball discharge traveling over the photographic plate suggests that here also the negative particles are the active ones.

The few who have search-lights have of late been throwing them upon the great mass of experimental work on the discharge through gases, published during the last generation. It is most instructive to remember that the Crookes tube was known for seventeen years before Roentgen discovered that something was going on outside of it. A repetition of some of the work done on spark discharge, and in particular the work of Wheatstone, in the light of what is now known, would be likely to yield results of the greatest value. It would be of particular value to study by the Wheatstone method the unipolar discharges of the Holtz machine.

A few words only may be added respecting radioactive phenomena.

We have long been familiar with the changes in matter, of a character such as may perhaps be described as spontaneous. Many crystals slowly lose their water of crystallization. They give off emanations. They explode very slowly. Now emanations, like all other matters and things, have individual peculiarities which enable us to recognize them. The emanation from crystallized sodium carbonate is also given off by all animals and plants, and is evidently a very useful and widely diffused substance. There are many substances which go to pieces and give off energy. They explode. Many of them give off more energy per gram per second than any radioactive body, while radium gives off more energy per gram than any other body. The radium explosion also goes on at a lower temperature than that of any other body. It hardly seems to be warranted, to say that the action is the same at the temperature of freezing hydrogen as at ordinary temperatures, for it does not seem that any high degree of precision has been attained in such a measurement. And certainly it can hardly be claimed that we know what these radioactive bodies would do at a temperature below 16 degrees absolute.

Seven years ago, an attempt was made in my laboratory to obtain

X-ray effects from explosions. High-grade gunpowder was loaded by strong compression into rifle-shells designed for a 40-grain charge. This powder was discharged from a heavy rifle, against an oak target six inches from the muzzle of the gun. The target was faced with thin plates of aluminum, which required frequent renewal. The concussion was sufficient to extinguish a gas-flame seven feet from the line of discharge. The plate used was one which would yield distinct X-ray effects from an exposure of one second to a Crookes tube, operated by an eight-plate Holtz machine. The dry plate was placed behind the target, and was subjected to the discharge of twenty-five pounds of powder, the operation requiring the spare time of the experimenter for forty days. The result was negative. No fluorescent effects could be detected by an observer behind the target. A rapid-fire gun might yield different results.

The same experiment was made with a thousand copper shells loaded with mercury fulminate. They were exploded in twos, one being fired electrically, the other being exploded by the concussion. The first shell was laid upon a wooden block resting on a two-inch plank. The second shell, to be exploded by it, was laid upon it with a heavy iron bolt-head just above. No metal was interposed between the explosive and the photographic film beneath the plank, and it was necessary to replace the block by a fresh one at each explosion. These explosions were so violent that a photographic plate of glass was shattered by the shock at almost every shot, and the windows thirty feet distant were perforated by bits of copper which occasionally escaped through the surrounding screens. A sensitive film of gelatine was used, on which the shadow picture was expected, but none was obtained. There is yet some reason to expect positive results from experiments of this kind. It may well be that explosives differ in this respect as in others. An investigation of the products of such explosions by the electrical means now used in the study of radioactive bodies is a wide and most inviting field, which is likely to aid in the explanation of radioactive phenomena.

Some of the products of explosion in the case of radium and uranium are more nearly elementary in character than other bodies yield, and some of the products are more elementary than others.

Now there is nothing unusual in finding here and there a substance which has some property to a very exalted degree. The diamond is such a case. Iron is vastly more magnetic than any other substance. All substances are magnetic. A group consisting of iron, cobalt, nickel, etc., are more magnetic than the great body of substances, and iron heads the list. There is nothing more remarkable in finding a group of radioactive substances with one which enormously surpasses all others than there is in finding an Academy of Science with some member surpassing all the others in some particular direction.

The relations which have been found to exist between atoms and molecules are no more disturbed by the behavior of radioactive substances than by the explosion of nitroglycerine. We have learned that what we have provisionally called atoms are, at least in some cases, as has long been believed, very complex in their structure. We should hardly expect an architect to lose confidence in houses, if he finally learns that the bricks with which he is familiar are not the final elements in their structure. That the bricks are made up of molecules, and the molecules of atoms, and the atoms of electrons, and that some houses have been observed to fall into pieces and give off energy, would hardly affect the usefulness of houses which do not fall to pieces, even if inertia is shown to be an electromagnetic phenomenon. And I think we should all remember that the proposition that matter has mass is fundamentally different from the proposition that a mass of matter has inertia. If inertia can be explained to be an electromagnetic quantity, and if it can be measured in new units, we have not changed the properties of matter. It is still matter, and it still has both mass and inertia. If inertia is an electromagnetic phenomenon, it may be measured in terms of the fundamental units in which all electromagnetic quantities are measured, — the units of length, time, and mass.

Formerly a force was measured in terms of the unit of mass only. People talked about a force of one pound. Later it was discovered that a force could also be measured in terms of the pound, the foot, and the second. At this time we did not hear any intimation that matter had had its day and was about to be abolished.

In physics we now think we have reached the domain of small things. But the electron may also be a very complex structure. If we accept Poynting's view of the nature of electromagnetic induction, the electron in a conductor is acted upon by a distant and moving electron, through a medium external to the conductor. The experimental verification of this is very convincing. In addition to this complex machinery we have to deal with machinery of gravitation.

We may always assume that nature is everywhere complex and ingenious. A visitor to our solar system, who should begin to study it from our earth, might begin with physical astronomy. He finally comes to chemistry, to zoölogy, and the phenomena of life, to governmental organization, to the moral and religious influences which dominate the lives and actions of men, to the simultaneous jurisdiction of state and federal courts within the same territory. By the time he had come to know this world as we know it, he would conclude that this universe of ours, which he first perceived as a faint and distant speck of light in the blazing firmament of stars, is, after all, very wonderful, and very much more complex than was at first believed. In arriving at our present ideas of the mechanism through which matter reacts on matter, we have not reached them by finding that the old ideas must be renounced, in order to explain some new phenomenon which is apparently out of harmony with the explanation previously made. It is rather that each new development has confirmed what had gone before, has made it seem more reasonable, and has filled in some gap in the knowledge of the past. The ether, which only a few years since was assumed to exist because it seemed to be necessary, has become more and more centrally important, and has finally come to monopolize most of the attention of those who would seek to understand matter. It is no reproach to modern ideas concerning the physics of matter, that they are complex. The fact that they are also harmonious and beautiful, and that they furnish an explanation of why a mass of matter has inertia, and promise the explanation of other long-standing puzzles, converts the accusation of complexity into a crowning glory.

Dewitt Bristol Brace
1859–1905

SECTION B — PHYSICS OF ETHER

(*Hall* 11, *September* 23, 3 *p. m.*)

CHAIRMAN: PROFESSOR HENRY CREW, Northwestern University.
SPEAKER: PROFESSOR DEWITT B. BRACE, University of Nebraska.
SECRETARY: PROFESSOR AUGUSTUS TROWBRIDGE, University of Wisconsin.

THE ETHER AND MOVING MATTER

BY DEWITT BRISTOL BRACE

[Dewitt Bristol Brace, Professor of Physics, University of Nebraska. b. Wilson, New York, 1859; died, October 2, 1905. A.B. Boston, 1881; A.M. Boston, 1882; Ph.D. Berlin, 1885; post-graduate, Massachusetts Institute of Technology, 1879–81; Johns Hopkins University, 1881–83; University of Berlin, 1883–85. Acting Assistant Professor of Physics, University of Michigan, 1886. Vice-President of American Association for the Advancement of Science; Vice-President of American Physical Society. **Author of** *Radiation and Absorption*.]

THE question whether the luminiferous ether passes freely through matter or participates in the translation of the same, considered as a moving system, stands to-day without positive answer, notwithstanding the numerous experimental attempts and the varied hypotheses which have been made since the discovery of aberration by Bradley in 1726. The simple explanation of this phenomenon on the corpuscular theory may have caused the century of delay in the closer examination of the question until it became necessary to consider it from the standpoint of undulations in an ether. As compared with the many efforts to examine the question in the second or ether period we have perhaps but two belonging to the first or corpuscular period. Boscovich, in 1742, reasoning from this theory on the ground of a difference of velocity in air and water, proposed to examine the aberration of a star with a telescope whose tube was filled with water. This experiment was not carried out till long after by Airy in 1872, who found that the variation in the aberration was absolutely insensible. Arago, in the second instance, reasoning on the same theory, concluded that the deviation produced by a prism would vary with the direction of the earth's motion; but he was unable to detect any such change, a result verified later by more delicate means in the hands of Maxwell, Mascart, and others. This experiment, which demonstrated the absence of any effect of the earth's movement on refraction is of great historical interest. This negative result, which to Arago was inconsistent with the corpuscular theory, suggested to Fresnel the important hypothesis of a quiescent ether penetrating the earth freely but undergoing a change

181

of density within the medium proportional to the square of its index and being convected in proportion to this excess of density, which would give an apparent velocity to the ether of $(1 - \mu^{-2})\nu$, instead of the velocity of the earth. Stokes suggested, as a simpler idea, that we suppose the ether is not convected but passes freely through the earth, being condensed as it passes into a body in the ratio of 1 to μ^2, so that its velocity within the refracting medium becomes $(1 - \mu^{-2})\nu$, from the law of continuity.[1]

Babinet in the second-century period attempted to test Fresnel's theory by examining the interference of two rays traversing a piece of glass, the one in the direction of the earth's motion and the other in the opposite direction. Stokes showed that a negative result was not contrary to the theory of aberration, since the retardation would be the same as if the earth were at rest.

He showed further, what Fresnel had not proven to be true in general, that on Fresnel's theory the laws of reflection and refraction for single refracting media are uninfluenced by the motion of the earth. In fact, Rayleigh has shown that, in using terrestrial sources, no optical effect can be produced by any system of reflecting or refracting optical surfaces moving as a rigidly connected system relatively to the ether, if we take into account the Döppler "effect," and neglect quantities of the second order of the aberration. Since, as Stokes says, the theory of a quiescent ether may be dispensed with, and as there is no good evidence that the ether moves quite freely through the solid mass of the earth, he proposes to explain the phenomenon of aberration on the undulation theory of light, upon the supposition that the earth and the planets carry a portion of the ether along with them, so that the ether close to their surfaces is at rest relatively to those surfaces and diminishes in velocity till at no great distance in space there is no motion. Cauchy had previously discussed the theory of a mobile ether, and had proposed to explain aberration by a shearing of the wave-fronts due to the translatory motion of the medium, but he did not develop his method sufficiently to explain how much the aberration would be.

On the other hand Stokes has specifically indicated his assumptions and formulated his conclusions. He examines the displacements of a wave-front in its passage from the ether at rest, across the region of transition to the ether in the neighborhood of the observer, which is at rest relatively to him. Adopting the same method which is used in the case of an ether at rest in determining the wave-front at any future time from that of a given one at any instant, he shows,

[1] If x is the velocity of the ether relative to the moving matter, and the density of ether within it is μ^2, the density of free ether being unity, we have from the law of continuity $\nu = (\nu - x)\mu^2$ and hence,

$$x \quad \frac{\nu\mu^2 - \nu}{\mu^2} = (1 - \mu^{-2})\nu$$

on the one condition, viz. that the motion of the ether is differentially irrotational, that if we neglect the square of the aberration and of the time, the change in direction of the ray as it travels along is *nil*, and therefore the course of a ray is a straight line, notwithstanding the motion of the ether. Following out the analysis on this supposition, a body, a star for example, will appear displaced toward the direction in which the earth is moving through an angle equal to the ratio of the velocity of the earth to that of light, when moving normal to the star's direction. This rectilinearity of propagation of a ray, which would likely seem to be interfered with in the motion of the ether, is the tacit assumption made in explaining aberration. If the physical causes, in consequence of which the motion of the ether becomes irrotational, could be adduced, the theory of Stokes would satisfy completely aberration and the negative results of the many and various experimental investigations which have thus far been made and whose validity is unquestioned, whether in refraction, interference, diffraction, rotary polarization, double refraction, induction, electric convection, etc. In an ordinary fluid, tangential forces proportional to the relative velocities destroy the irrotational condition in a steady state of motion. If we suppose these forces to be diminished indefinitely we obtain now a motion totally different from that for the steady state when these forces are assumed to be absent initially; and hence such a motion would be unstable. When, however, tangential forces depending on relative displacements in the ether are considered, it becomes possible to explain the irrotational condition. Any deviation from this state, for example at a surface of *slip*, would be dissipated away into space with the velocity of light by means of transverse vibrations. He illustrates such apparent incompatibilities in physical states by successive dilutions of gelatine. Such a medium shows elastic tangential forces for small constraints, and yet apparent fluidity for motions through it, mending itself as soon as dislocated. He regards these qualities as consistent and self-sufficient to explain the phenomena in question. Against the view of Stokes, Lorentz raises objection to his assumptions concerning the ether motions in the neighborhood of the earth, which he considers inconsistent, a difficulty which he is unable to set aside. Larmor demurs against an appeal to a highly complex medium, such as pitch, for studying the behavior of a simple one like the ether. A time-rate much shorter than the time of relaxation will of course provide approximate rigidity, while a time-rate much longer will provide approximate fluidity, but this requires inevitable dissipation. This objection would be valid for a viscous solid, but such Stokes apparently did not have in mind, since he specifically proves such a case unstable. A solid like pitch is a very different type of solid from that of a vesicular solid like jelly. An ether after the model of a viscous solid would always contain the viscous terms, so that even for the high time-rates of light-waves there would be dissipation however small. Such a condition, it can be proven, would give coloration to the remote members of the stellar system; a fact inconsistent with observation. On the other hand, a soft vesicular solid like gelatine may not necessarily contain the time-factor, and yet be so soft that dislocation may occur even with constraints of the order of aberration, but not of the square of that order. Such an ether without a *time of relaxation* factor would fulfill completely the conditions of a luminiferous ether, if, as Stokes tried to show, it could be reconciled with the phenomena of aberration and the motions of the heavenly bodies. The method of double refraction shows that a solution of gelatine of one part in a thousand is rigid, while at the same time it appears as mobile as water, and its rate of flow through small tubes does not vary largely from the same. This experiment illustrates very markedly Stokes's example. When such a solution is continuously dislocated between two surfaces in relative motion, the same double refraction is present, indicating that the stress is still active during dislocation. Also a metal, like copper, shows a similar stress while being strained beyond its elastic limit. If this takes place by *slip* or dislocation throughout the mass which, though irregular, may give a mean uniformity for sensible dimensions, such a medium might serve as our model. Any deviation from perfect regularity in molecular distribution and activity we might anticipate would give such minute irregular dislocations at the limit of elasticity. Such a medium would thus transmit completely any disturbance within this strain limit.

It is difficult, however, to conceive of the transmissions of a disturbance across a surface of dislocation. For many ordinary media, we should expect at such a surface total reflection. If we suppose such a transmission of disturbance, its mode is not apparent, even if we suppose a thin lamina in rotational motion which would diffuse at least a portion, if not all, of the incident disturbance. Similar difficulties would arise if we assume the ether a solid which becomes fluid under stress and thus allows bodies to pass through it (as, for example, through a block of ice, as Fitzgerald suggested). While such solutions may seem highly artificial and do violence to our convictions, the consequences of a quiescent ether may, when fully developed and tested, demonstrate its impossibility and command a more extended examination into the structural qualities of an all-sufficient medium than the single case of an essentially vesicular medium like jelly brought forward by Stokes and in a different form as a contractile ether by Kelvin. The theory of Fresnel of a quiescent ether in space presupposes a change of its density proportional to μ^2 within a ponderable medium, and a convection coefficient

$(1 - \mu^2)\nu$. This hypothesis satisfies the phenomena of aberration and the uniformity of the laws of reflection and refraction of a body, whether in motion or at rest, and, as already mentioned, does not affect interference, as Stokes showed, so far as the earth's motion is concerned. That the ether apparently is carried along within moving matter not with its full velocity, but diminished to the extent indicated by Fresnel's coefficient of convection, Fizeau demonstrated in his famous interference experiment with streaming water, repeated later with greater refinement by Michelson and Morley. The significance of this experiment in its bearings on the question of the drift of the ether has perhaps been overestimated. In fact, neglecting the square of the aberration, it is exactly what we should expect from the dynamical reaction of a moving material system on a periodic disturbance, propagated through it without reference to the motion of translation of the interpenetrating medium, but simply to the frequency of the vibration impressed upon the system by this ether. Thus if we transform the ordinary differential equations of motion of the material system from fixed to moving axes, the form of the solution contains Fresnel's convection coefficient as a factor exactly, neglecting quantities of the second order of the aberration. This experiment cannot then be adduced as a positive result in favor of a quiescent ether. On account of its physical consequences, however, it should be extended to the case of gases and to absorbing substances, using light corresponding to the natural frequences of the latter if possible. Although negative results have heretofore been obtained with a gas, yet, with high pressures and greater dimensions and velocities, the test is within present experimental limitations. Results with solid bodies are still lacking, but a preliminary examination of the problem encourages us to expect successful results, at least with double-refracting substances. Reasoning in a similar manner as on the dynamical reaction of a moving system, we should look for the acceleration of a circularly polarized ray propagated coaxially within a rapidly rotating medium. This may possibly be brought within experimental limits. Again we have the important experiment of Lodge on the effect of moving masses upon the motion of the ether near them. This experiment, like that of the preceding one of Fizeau, is a first order test, $i.\ e.$ the effect to be observed would arise from a change in the first power of the aberration factor. Two interfering beams were sent around several times in opposite directions between two rotating steel disks, and the effect on the bands noted from rest to motion or reversal. With a linear velocity not far from one two-hundredth that of the earth's orbital motion, and a distance of some ten meters or more, no influence on the interfering rays could be detected, thus making the effect, calculated from the aberration factor if the ether were carried around between the disks, something like twice the limit of observation. Lodge estimates from this experiment that the disks must have communicated less than the eight-hundredth part of their velocity to the ether. It is to be noted that the masses of these disks were not great, being only some two or three centimeters thick and about one meter in diameter. If we suppose the ether to be set in motion by means of reactions of a viscous nature, the experiment would be conclusive. To this extent, that the ether is not viscous, the test seems to be valid, but as there are other modes conceivable by which such movement might be brought about, it is not conclusive. If now we have to give up the notion of a quiescent ether, it will be necessary to suppose such motions are engendered in some way depending on the mass of the moving system, which we might imagine to be the fact in the case of the earth and the surrounding ether (possibly as, Des Coudres suggests, through gravitational action). It would be desirable to repeat this experiment, using great masses, and also testing to a much higher degree of sensibility (the third order would be possible) by means of double refraction. Michelson has recently attempted to determine directly whether the velocity of the ether diminished as we recede from the earth, but with negative results. He sent two interfering rays in opposite directions around the four sides of a rectangle of iron piping from which the air had been exhausted, the same being in a vertical east and west plane, the horizontal length of which was 200 feet and the height 50 feet. Assuming an exponential law for the variation in the velocity of the ether as we recede from the earth, he finds that if the earth carries the ether with it, this influence must extend to a distance comparable with the earth's diameter. The negative result in many of the experiments on refraction and interference which different investigators have obtained and which apparently follow on the assumption of a mobile ether have been usually experiments capable of giving only second order effects instead of the first order effects looked for, which, as mentioned above, are quite as consistent with a quiescent ether, as Stokes and Rayleigh have shown. Among these may be mentioned the experiments of Hoek, Ketteler, Mascart, and others on interference in ponderable media, over opposite paths relatively to the earth's motion; as also those of the two latter with double-refracting media. All of the experiments were first order tests, and hence should give negative results on either theory, since, with a terrestrial source of light, the phenomena are independent of the orientation of the apparatus neglecting second order effects.

The positive results of Fizeau and of Ångström have not been confirmed and should not be seriously considered. In the experiments of the latter, the variation of the position of the Frauenhofer lines, as obtained by a grating when observed in directions with and opposite to the earth's orbital motion, has never been noted since, beyond the anticipated displacement calculated from the purely kinetical principle of Döppler. The experiments of the former, as a first order test, on the rotation of the plane of polarization of a ray after passing through a pile of plates has perhaps offered the greatest difficulty to the exponents of both theories in reconciling the observations with the results which should follow from each theory. In this experiment, performed in 1859, the optical systems was mounted so as to be rotated about a vertical axis alternately from east to west, or *vice versa*. This system consisted of the usual polarizing nicol or sensitive tint-system and analyzing nicol between which were placed several piles of plates and compensating systems for producing the rotations and the magnifying of the same, and also for compensating for the rotary dispersion and elliptic polarization of the transmitted light which was polarized in an azimuth of 45°. In a series of observations extending over some time the mean of the rotations of the plane of polarization showed a maximum excess in the direction toward the west at noon and at the time of the solstice. It is to be noted that light from a heliostat was reflected into the system alternately by two fixed mirrors when the system was rotated. This required an interruption and readjustment of the heliostat during a single observation, *i. e.* from east to west and west to east, the difference in the setting of the analyzer in the two positions to give the same field of view being, of course, the effect sought for. Fizeau refers to the irregularities arising from successive settings of the heliostat. The calculated effect was much below that which could have been observed directly with the usual polarizing system. To magnify any such effect, a second system of plates was used which gave an amplification as high as eighty times. Thus any residual rotation from whatever cause would receive the corresponding amplifications. Now, in experiments with polarizing systems using sunlight as a source of illumination, it has frequently been noted that any shift in the direction of the light through the apparatus, either due to a change in the direction of the beam (arriving, say, from the heliostat) or to a shift in the optical system itself, produced a change in the field of view, whether with a half-shade system or otherwise. In the former the match was destroyed, the change being of an order much greater than that which Fizeau anticipated from calculation. Further, with such limited beams of light, a mere shift of the eye may produce an effect of similar magnitude. Hence, in all polariscopic experiments where sunlight is used, it is absolutely essential that, during any single observation, the ray of light pass through the system and into the eye over exactly the same path. This Fizeau failed to carry out, and this is entirely sufficient to explain the very great discrepancies in his various series of observations, and probably the apparent constant difference in the results of his settings in the two directions.

In fact, Fizeau himself has stated since that his observations were not absolutely decisive. While the test is now probably within experimental limits with the more highly refined half-shade systems, other modes of experimenting on different optical principles with greater sensibilities have given negative results, thus disproving the existence of a phenomenon which Fizeau's experiment apparently established, and making a repetition of this experiment, which is of doubtful execution, unnecessary.

The effect of the motion of a natural rotative substance through the ether on the rotation of the plane of polarization is of considerable importance in its bearings on certain controverted points in some of the recent theories of a quiescent ether. Mascart, who first studied the problem in the case of quartz, was unable to detect any difference in the rotation when a ray was propagated in and against the direction of motion of the earth. This variation in the total rotation, which he could detect, was one part in 20,000, or one part in 40,000 on reversal. This experiment as thus carried out corresponds to a first order effect. Rayleigh quite recently has repeated this experiment with a sensibility five times as great, and obtained negative results, likewise. The impossibility of obtaining quartz in sufficient quantity and purity, or natural rotary liquids of sufficient power, to attain the extreme limit of polariscopic possibilities seems to make even an approximation to a second order effect entirely improbable, although the higher frequencies might be used, where the power may be ten times as great. On the other hand, the effect of the mechanical rotation of such a medium on the circular components is, however, probably not beyond experimental possibilities in polariscopic work.

On the electrical side several first order experiments have been made which likewise have given negative results. Des Coudres has attempted to determine the difference in the induction on each of two coils placed symmetrically, with respect to a third coaxial coil between them. On compensating for the effects of each on the galvanometer when the axis of the system was in the direction of drift, and then reversing the direction of the system, no influence on the galvanometer could be observed. The effect which should be observed corresponds to the second order of the aberration. However, without compensating factors, the theory of induction phenomena shows that second order effects should be looked for in systems moving through the ether. The same may be said of other electrical experiments.

The difficulties in formulating a theory which will explain the results of all experiments involving tests to the first order of sensibility only on the assumption of either a quiescent or a convected ether, are much easier met than when second and higher orders have to be taken into consideration. Here we find what, at first sight, appear as rather startling assumptions; but it is only in this manner that present observational facts can be reconciled with a quiescent ether. With each advance in experimental refinement, theory has had to adapt itself by the adoption of new hypotheses. This has now been done up to second order phenomena for a quiescent ether. Thus far, however, no hypothesis has been brought forward to adapt specifically the theory of a quiescent ether to observations which have already been carried up to the third order of the aberration constant.

The first second order experiment was carried out by Michelson and Morley, and was an optical test in which the method of interference of two rays passing over paths mutually at right angles to one another was used. The apparent intent of the originators of this experiment was initially to look for a first order change in the aberration factor by means of a second order interference effect. The difficulty in reconciling the negative results of this test has, however, given rise to hypotheses involving second order dimensional factors, so that from this point of view it becomes a second order experiment. It could not, however, show a first order change in the velocity of the moving system, which latter, referred to the velocity of light, is taken as a magnitude of the first order, and hence the former change would count as a second order magnitude. In this experiment the entire system was mounted on a float so that the optical system could be rotated consecutively through all quadrants of the circle while the interference bands were being continuously observed. If now the difference in time of passage of one of the rays, say along the line of drift, and the other at right angles to it, is calculated on the basis of a moving ether, we find it to be equivalent to the time of passage over a length corresponding to a diminution of this length, in the direction of drift, proportional to the square of the aberration. Their results show that had there been an effect, it must have been probably sixteen times, certainly eight times, less than that calculated. It is understood that Morley and Miller will soon report as the result of a repetition, during the present year, of this experiment on a much larger scale, that, if there is any effect, it must be one hundred times less than the calculated value. This result is entirely consistent with a moving ether, but seemingly contradictory to a quiescent ether, as proposed by Fresnel. Apparently, then, either some condition in the fundamental hypothesis of such a medium has been overlooked, or a supplementary hypothesis must be imagined. Similar hypotheses were conceived of by both Lorentz and Fitzgerald independently, shortly after the publication of the experiments of Michelson and Morley in 1887. They assume that a contraction in the direction of motion takes place in a system moving through the ether, so that this dimension is reduced by a fraction of itself equal to one half the square of the constant of aberration. This of course, as an assumption, merely suggests a compensation to meet an apparent residual effect, and would be of no significance if it were impossible to incorporate such a condition into a consistent theory of ethereal action. This has been done by Lorentz and by Larmor in their theories of moving systems. Lorentz, who was the first to develop a satisfactory theory of a quiescent ether, assumes that, in all electrical and optical phenomena taking place in ponderable matter, we have to deal with charged particles, free to move in conductors, but confined in dielectrics to definite positions of equilibrium. These particles are perfectly permeable to the ether, so that they can move while the ether remains at rest.

If now we apply the ordinary electromagnetic equations of a system of bodies at rest to a system having a constant velocity of translation in addition to the velocities of its elements, the ether remaining at rest, the displacements of the electrons arising from the electric vibrations in the ether and the electric and magnetic forces are the same functions of the new system of parameters as for the case of rest, if we neglect quantities of the second order of the aberration. This theorem assumes that the distance of molecular action is confined to such excessively small distances that the difference in their local times would have no effect. An exception to this may be found in a rotary substance like quartz which, as mentioned above, has been examined by Mascart and Rayleigh to the first order with negative results, which seems to warrant the conclusion that the molecular forces are themselves altered by translation. This theory of Lorentz seems capable, then, of explaining the uniformly negative results of all the first order tests which have been described previously, without, however, necessarily establishing it finally, since we have not yet studied its adaptability to second and higher orders of the aberration.

The suggestion of a contraction, as stated above, lends itself in a similar manner and under like restrictions to that for the first order transformation. This requires the introduction of a second coefficient differing from unity by a quantity of the second order as did the coefficient used in the first transformation, but differing from the latter in that it is left indeterminate from the fact that there are no means as yet for giving it a definite value. Introducing these new parameters we again obtain a set of equations in which the velocity of translation does not explicitly appear. Such a moving system has therefore its correlate in a system at rest, the former having changed into the latter through the assumed contraction the moment motion begins. The occurrence of these coefficients as factors in the electric forces and the accelerations arising from the electric vibrations in the ether in the expression for the corresponding system at rest, necessitates that if the degree of similarity required is to exist in the two systems, the electrons must have different masses depending on whether their vibrations are parallel or perpendicular to the velocity of translation. This startling conclusion of Lorentz is borne out by what we now know of the dependence of the effective mass of an electron upon what is taking place in the ether. Such an hypothesis as this would require that Michelson and Morley's experiment should always give a negative result.

Of electrical experiments on the drift of the ether we have one second order test carried out very recently by Trouton at the suggestion of the late Professor Fitzgerald. The latter, reasoning on the condition of a magnetic field produced by a charged condenser moving edgewise to the drift of the ether, and the consequent additional supply of energy of such a system on charging, thought that this might produce a mechanical drag on charging and an opposite impulse on discharging, just as might occur if the mass of earth were to become suddenly greater. This experiment was carried out in the form of a condenser mounted upon an arm carried by a delicate suspension, with negative results. A second and more sensitive test was made later in a modified form by Trouton and Noble. Since. edge on to the drift, we have a magnetic field, while at right angles it vanishes, the energy will vary with the azimuth, and we shall have a maximum in an azimuth of 45°. A delicate suspension carrying the armature of a condenser showed no movement, although the calculated effect was ten times the limit of observation. The negative results of these experiments may be accounted for on like assumptions with that of the Michelson and Morley experiment, namely a contraction or change in the dimensions of the condenser producing corresponding changes in density and potential difference of the charge.

The assumption of a contraction suggests at once, from what we know of transparent media, the anisotropic state which such media are thrown into under dimensional strain. Rayleigh has examined this question in the case of water, carbon disulphide, and glass without result. In the case of glass his sensibility was several times the calculated second order effect, and much more in case of liquids.

The degree of refinement to which the polariscopic test lends itself is perhaps beyond that of any other instance in physical application. Here then is an opportunity to examine the question beyond what theory has anticipated, and the test has been carried so as to reach safely a third order effect, with negative results. The experiments as performed by the writer consisted in sending a beam of sunlight plane polarized at 45° to the horizon, through 28.56 meters of water in a horizontal direction and examining the same by a sensitive elliptic analyzer. On rotating the entire system from the meridian, where the one component of vibration to the drift was parallel

191

and the other perpendicular, into a plane at right angles to the meridian where both components would be at right angles to the drift, and therefore where no differential effect would be produced, no change in the field of view could be detected. Had there been a total difference of 7.8×10^{-13} of the whole velocity between the components, the effect would have been manifest. We may, therefore, conclude that there is no third order effect. How well the various theories of a quiescent ether will lend themselves to this further adaptation remains to be seen, but undoubtedly by properly choosing the coefficients it may be done; however, any theory which does not contain explicitly the exact and complete adaptation to all orders of the aberration must certainly impress itself as highly artificial in its successive auxiliary hypotheses and approximations.

Larmor, in reference to his theory, says, "It is, in fact, found that the Maxwellian circuital equations of æthereal activity, in the ambient æther referred to axes moving along with the uniform velocity of convection, v, can be reduced to the same form as for axes at rest up to and including $\left(\dfrac{v}{V}\right)^2$ but not $\left(\dfrac{v}{V}\right)^3$ by adopting certain coefficients." "If, then, matter is for physical purposes a purely æthereal system, if it is constituted of simple polar singularities or electrons, positive and negative, in the Maxwellian æther, the nuclei of which may be either practically points or else small regions of æther with internal connections of pure constraint, the propositions above stated for the first order are extended to the second order of $\dfrac{v}{V}$ with the single addition of the Fitzgerald-Lorentz shrinkage in the scale of space and an equal one in the scale of time, which, being isotropic, is unrecognizable." "On such a theory as this the criticism presents itself, and was in fact at once made, that one hypothesis is needed to annul optical effects to the first order; that when these were found to be actually null to the second order, another hypothesis had to be added: and that another hypothesis would be required for the third order, while in fact there was no reason to believe that they were not exactly null to all orders. Such a train of remarks indicates that the nature of the hypothesis has been overlooked. And if indeed it could be proved that the optical effect is null up to the third order, that circumstance would not demolish the theory, but would rather point to some finer adjustment than it provides for; needless to say the attempt would indefinitely transcend existing experimental possibilities." And further, "up to the first order the electron hypothesis, that electricity is atomic, suffices by itself, as Lorentz was first to show." "Up to the second order, the hypothesis that matter is constituted electrically — of electrons — is required in addition."

The necessity in view of the present experimental data for leaving indeterminate the units of transformation is here illustrated in the theory of Larmor.

In the most recent discussion by Lorentz, the necessity of a general treatment is shown for not only the second but also the higher orders. In a consideration of transparent media, his theory attempts to show that translation would not alter interference, diffraction, or polarization. He would thus, by means of the assumption of so-called "Heaviside ellipsoids" as the shape of electrons, explain the negative results of optical experiments, as well as the observations of Kaufmann on Becquerel rays.

Attention should also be called to the recent theory of Abraham, who gives as the ratio of the axes of the moving electron $1 - \frac{4}{5}\left(\frac{v}{V}\right)^3 : 1$, omitting fourth and higher orders. This would give a residual in double refraction of $\frac{1}{5}\left(\frac{v}{V}\right)^2 = 2 \times 10^{-9}$ for transparent media, which he acknowledges is difficult to reconcile with the experimental results which show no double refraction to the first order beyond this.

Paul Langevin
1872–1946

SECTION C—PHYSICS OF THE ELECTRON

(*Hall 5, September 22, 3 p. m.*)

CHAIRMAN: PROFESSOR A. G. WEBSTER, Clark University.
SPEAKERS: PROFESSOR PAUL LANGEVIN, Collège de France.
  PROFESSOR ERNEST RUTHERFORD, McGill University, Montreal.
SECRETARY: PROFESSOR W. J. HUMPHREYS, Mount Weather, Va.

THE RELATIONS OF PHYSICS OF ELECTRONS TO OTHER BRANCHES OF SCIENCE

BY PAUL LANGEVIN

(*Translated from the French by Bergen Davis, Ph.D., Columbia University*)

[Paul Langevin, Assistant Professor of Physics, Collège de France, Paris, since 1903. b. Paris, France, January 23, 1872. Licencié in Physical and Mathematical Science; Fellow of the University; Ph.S.D.; Instructor at Sorbonne, 1899–1903.]

THE remarkable fertility shown by the new idea, based on the experimental fact of the discontinuous corpuscular structure of electrical charges, appears to be the most striking characteristic of the recent progress in electricity.

The consequences extend through all parts of the old physics; especially in electromagnetism, in optics, in radiant heat; they throw a new light even on the fundamental ideas of the Newtonian mechanics, and have revived the old atomistic ideas and caused them to be lifted from the rank of hypotheses to that of principles, owing to the proper relation which the laws of electrolysis have established between the discontinuous structure of matter and that of electricity.

Without seeking here to run through the whole field of their applications, I hope to indicate upon what solid foundations, both experimental and theoretical, rests at present the notion of the electron so fundamental to the new physics; to indicate the points which seem to require more complete light, and to show how vast is the synthesis which we can hope to attain, a synthesis whose main lines only are fixed to-day.

Under actual and provisional form, this synthesis constitutes an admirable instrument of research, and owing to it the questions extend in all directions. There is there a kind of New America, full of wealth yet unknown, where one can breathe freely, which invites all our activities, and which can teach many things to the Old World.

195

I. *The Electromagnetic Ether*

(1) *Fields and Charges.* One can say that the combined efforts of Faraday, Maxwell, and Hertz have resulted in giving us a precise knowledge of the properties of the electromagnetic ether, and of light; of a medium, homogeneous and void of matter, whose state is completely defined, with the exception of gravitation, when we know at any point the direction and magnitude of the electric and magnetic fields.

I insist, for the present, on the possibility of arriving at a conception of fields of force, as well as the related idea of electric charges, independently of all dynamics; I wish by this to imply only a knowledge of the laws of motion and of matter.

The two fields possess this property, that their divergence is zero in all parts of the ether; that is to say, the flux of electric and magnetic force is rigorously zero across a closed surface which does not contain any matter in its interior. It is in fact always matter in the ordinary sense of the word which contains and can furnish the electric charges around which the divergence of field exists whose direction varies with the sign of the charges.

In extreme cases where the electric charges appear to be most completely separated from their material support, as in the case of the cathode rays for example, the experimental fact of the granular structure of these rays and the complete indestructibility of their charge, the fact finally that cathodic particles are charges possessing the fundamental property of matter, inertia, and experiencing acceleration in the electromagnetic field, these facts do not allow us to distinguish their charge from the so-called free charge of ordinary electrified matter.

Furthermore, we shall come to the idea not only that there can be no electric charge without matter, but that, in fact, there can be no matter without electricity, an aggregation of electrical centres of the two kinds. Electrons, analogous to the cathode particles, possess almost all the known properties of matter by the fact alone that these centres are electrified. We shall see within what limits this conception can be considered sufficiently known, and if it is necessary to superimpose other properties on those which result from electrically charged centres in order to obtain a satisfactory representation of matter; the ether alone, on the contrary, never contains any electricity.

If experiment obliges us to admit the existence of electric charges, positive and negative, from the flux of electric force different from zero across a closed surface drawn entirely in the ether and containing matter, it is otherwise for the magnetic field. Experiment has never furnished an instance where a closed surface drawn in the ether was traversed by a magnetic field different from zero. One interesting phenomenon observed recently by P. Villard in the effect of an intense magnetic field on the production of the cathode rays, appears to receive a simple explanation in the hypothesis of free magnetic charges; but it is not certain that this hypothesis is necessary.

(2) *The Equations of Hertz*. The two fields, electric and magnetic, of which the ether can be the seat, are related to one another in such a manner that one of them can exist only on the condition that the other varies; all variations of an electric field produce a magnetic field; it is the displacement current of Maxwell: and all variations of the magnetic field produce an electric field; this is the phenomenon of induction discovered by Faraday. These two relations are expressed by Hertz's equations; they sum up completely our knowledge of the electromagnetic medium, and from these it results that all disturbances of this medium are propagated with the velocity of light. Hertz had the glory of proving this fact experimentally.

(3) *Energy*. We can now say that the ether is the seat of two distinct forms of energy, the electric and the magnetic, capable of transformation from the one into the other, *but only through matter as an intermediary, that is to say, by means of the electrified centres which it contains.*

In the ether alone, in fact, in the free radiation which it propagates, the electric and magnetic fields, transverse with respect to the direction of propagation, represent always equal energies in each element of volume, without oscillation of the energy from one form to the other. In the presence of matter, on the other hand, the electric energy can exist alone, and it is the motion of electrified centres which allows the transformation into magnetic energy, and *vice versa. Matter only can be the source of radiation.*

It is necessary, to the two preceding forms of energy, to add gravitation, which corresponds probably to a third mode of activity of the ether, whose connection with the two others is still obscure.

I insist here on the point that the principle of equivalence of various forms of energy, as far as the process allows of measurement, can be attained independently of all dynamical notions, by the process of using solely material systems in equilibrium.

One can find some information on this subject in a recent exposition by M. Perrin.[1]

(4) *The Theory of Lorentz*. The ether being thus completely known to us from the electromagnetic and optical point of view, the problem which follows as a continuation of the work of Maxwell and of Hertz is that of the connection between ether and matter, inert matter, the source and recipient of the radiations which the ether

[1] I. Perrin, *Traité de chimie Physique. Les Principes.* Gauthier-Villars, Paris.

transmits. The connection sought for is furnished us by the electron or corpuscle, an electrical centre movable with respect to the ether, and carrying with it its divergent electric field.

This was the fundamental idea which caused Lorentz to conceive of the possibility of a relative displacement of electrified centres of divergence of the electric field, and of the ether considered as immovable. This displacement takes place without any change in the amount of the charge, that is to say, that the surface which is displaced in the ether with the electron is crossed by an electric flux which is completely invariable. It is the fundamental principle of the conservation of electricity, which will perhaps absorb the principle of the conservation of matter, as we cannot have matter without electricity. It is, however, probable that electricity alone is not sufficient to constitute matter.

We have actually no very precise information of the relative displacement of charges and of the ether, of electrified centres in an immovable medium, no tangible form under which we can conceive it. The attempts which have thus far been made to obtain a concrete representation, in order to give a material structure to the ether, have all been sterile of results. Perhaps there is a difficulty which belongs to the actual constitution of our minds, habituated by our secular evolution to think through matter, unable to form a concrete representation which is not material; also it seems scarcely reasonable to seek to construct a simple medium such as the ether by considering it to spring from a complex and various medium like matter. I believe it will be necessary to think *ether*, to conceive of it independently of all material representations, by means of those electromagnetic properties which put us in contact with it. I will return to this point later in reference to the mechanical theories of the ether.

If the electric charge is assumed to have a volume distribution in a portion of the medium, the principle of the conservation of electricity, and also the possibility of relative displacement of electricity and ether, makes it necessary for us, in this portion of space, to modify the equations of Hertz relative to the displacement current by the addition of a convection current, a necessary consequence of the existence of a displacement current connected with a motion of charges, and implying the production of a magnetic field by the motion of electrified bodies across the medium. This consequence of Hertz's equations has now received complete experimental confirmation.

Moreover, the experimental facts impose on these movable charges a discontinuous, granular structure, and lead to the idea of the electron as a singular region of the ether, carrying a charge equal to that of the hydrogen atom in electrolysis, but of different sign, and distributed on the surface or in the volume of the electron according as the intensity of the electric field is supposed to present, or not, a discontinuity when it crosses the surface which limits the volume occupied by the electron. Inertia, of electromagnetic origin, which we are about to refer to a similar centre, is opposed also, under the difficulty of its becoming infinite, to the hypothesis of a finite electric charge condensed in a point without extension.

The various considerations, more and more precise, all converging toward this notion of the atomic structure of charges, form the starting-point of all recent works on electricity.

II. *The Atom of Electricity*

(6) *The Electron.* The remarkable laws of electrolysis discovered by Faraday establish an intimate and necessary connection between the atomic structure of matter and that of electricity. They were sufficient to lead Helmholtz to conceive the latter as constituted of distinct, indivisible portions, elements of charge, all identical from the point of view of the quantity of electricity which they carry, and differing only in the sign. This elementary charge is equal to that carried by a monovalent atom or radical in electrolysis; a polyvalent atom or radical carries an equivalent number of such charges.

It was Johnstone Stoney who first used the word electron to designate atoms of electricity as distinct from matter, with which they combine to furnish the electrolytic ions. The presence of similar electrons combined with material atoms allows us to represent certain peculiarities of the spectrum, the existence of doublets of like frequencies; the electron, in motion, is thus considered as the origin of the emission of all luminous rays.

(7) *Gaseous Conductors.* But there are the researches on the electrical conductivity of gases, which have presented to us in a forcible manner the idea of electrical atoms, which have made this notion more tangible by allowing us to count these electric centres, to lay hold of them individually, and to measure for the first time the charge of each of them in absolute value.

As early as 1882, Giese, in observing the peculiarities of the conductivity of gases escaping from flames, the departure from Ohm's law, the impossibility of drawing from the gas, whatever might be the electric field employed, more than a limited amount of electricity of each kind, the progressive recombination of the free charges in the gas, had expressed in a precise manner the idea, that as in electrolytes the free electric charges in a gas are carried by distinct positive and negative centres in limited numbers, capable of moving in opposite directions under the action of an external electric field in order to discharge the electrified body which produces the field.

It is difficult, in fact, to conceive how, on the hypothesis that the charges are distributed in a continuous manner in space, a mass of gas electrically neutral could furnish a limited quantity of electricity of each kind, decreasing with the time by progressive recombination if one delays the establishment of the electric field in the gas.

It is indeed necessary to admit, for the two electricities, a discontinuous structure in order to allow their coexistence without completely neutralizing one another. The progressive recombination of the charged particles or ions of two kinds would produce this neutralization at the moment of their mutual collisions.

The phenomena of the saturation current, of the limited quantity of free electricity in a gas, were obtained under conditions most favorable to experimental study, when, immediately after the discovery of Roentgen rays and like radiations, one had recognized their property of making the gas they traversed a conductor of electricity. The limited charge which we can extract from a gas thus modified, the velocity, finite and easily measured, with which they move under the action of an electric field, their progressive recombination, are interpreted in an admirable manner on the hypothesis that the radiations, as well as the intense heat agitations in a flame, dissociate a certain number of the molecules of the gas into electrified parts carrying charges of opposite kinds.

(8) *The Phenomena of Condensation.* We know how the phenomena of condensation of supersaturated water vapor in the presence of a conducting gas, already referred by R. von Helmholtz to the presence of ions, has given the preceding hypothesis a brilliant confirmation. As a result of the researches of J. J. Thomson, Townsend, C. T. R. Wilson, and H. A. Wilson, these droplets of visible water, each formed by condensation around an electrified centre, bring forward a tangible witness to the existence of these centres, and furnish a means of measuring the individual charge, present on each drop of water formed, and equal to about 3.4×10^{-10} electrostatic units of electricity according to the recent measurements of J. J. Thomson and H. A. Wilson.

The fundamental idea in these kinds of measurements, applied for the first time by Townsend to the charged drops which are produced in the presence of saturated water vapor in recently prepared gases, consists in deducing the mass of each drop from its velocity of fall under the action of gravity by means of Stokes's formula, which gives the frictional resistance of a sphere moving through a viscous medium, and which expresses the velocity of fall in terms of the radius of the drop and consequently of its mass. We can obtain from this the electric charge carried by each drop if we know the ratio of this charge to the mass.

This ratio can be obtained, as was done by Townsend and J. J. Thomson, by measuring or calculating the total mass of water carried by the droplets, considered as uniform, as well as the total quantity of electricity carried by the ions which have served as centres for the formation of the drops. The charge thus obtained by Townsend was found to be 3×10^{-10} electrostatic units for each centre in the case of gases of electrolysis, and to 6.5×10^{-10} by J. J. Thomson from the first series of measurement on gases ionized by Roentgen rays.

H. A. Wilson obtained the ratio of charge to the mass of a drop more simply by comparing the velocity of fall under the action of gravity alone with the velocity of fall in a vertical electric field. He obtained thus directly the ratio sought for. This method has the advantage of showing that the electric charges are really carried by the drops, and of separating those drops which carry a single elementary charge from those which, by diffusion of the ions toward one another, carry a double or triple charge.

Wilson gives as the mean result of his measurements 3.1×10^{-10}, a value very near to that of Townsend.

A second series of experiments by Professor J. J. Thomson, in which he used radioactive substances as sources of ionization more constant than the Crookes tube, and in which he took care to cause the drops to form on all the ions present in the gas, by producing a supersaturation of the water vapor by a rapid expansion of sufficient magnitude to cause the condensation on the ions of both kinds, gave as a mean result 3.4×10^{-10}, a value in complete agreement with the other two experimenters. The principles of thermodynamics account perfectly for the influence of electrified centres on the condensation of water vapor: the electric charge of a drop in fact diminishes the pressure of water vapor in equilibrium with it. Moreover, the least supersaturation found necessary, by C. T. R. Wilson, for the formation of drops of water on the ions, which are the same whatever may be the means of producing them (Roentgen rays, Becquerel rays, brush discharge, action of ultra-violet light on metal negatively charged), allows us by purely thermodynamical reasoning to calculate approximately the charge carried by each of the ions, and this calculation, entirely distinct from direct measurement, gives in the case of the positive centres a value of 4×10^{-10} E. S. units.

(9) *The Radiation Integral.* More surprising still is the result recently obtained by H. A. Lorentz, who succeeded in basing a precise measurement of the elementary charges carried by the electrified centres present in metals on the experimental study of the radiation integral or black body radiation.

We will see how the emission and absorption of heat- and light-waves by matter are dependent on the presence in it of electrons in motion. The ratio, for a radiation of given wave-length, between the emissive and absorptive power, a ratio independent of the nature of the substance, represents the emissive power of the radiation integral, which bolometric measurements give directly.

Now this ratio can be calculated, as Lorentz has shown, for wavelengths which are long in comparison with the mean path of free electrons in the metal, as a function of the charge carried by each of them. The comparison of these results with those of Kurlbaum furnishes an entirely new method of obtaining this charge, and gives 3.7×10^{-10} E. S. units.

(10) *The Kinetic Theory.* Finally, the last confirmation, which states more precisely still our knowledge of the electric atom, and our confidence in this fundamental idea, Townsend, through comparing by the simple reasoning of the kinetic theory the velocities of ions in a gas under the action of an electric field with their coefficient of diffusion through the interior of the gas, two quantities directly measurable by experiment, has been able to demonstrate the identity of the charge of one of these gaseous ions with the electric atom of Helmholtz, the charge of a monovalent atom in electrolysis.

From this comes directly a new confirmation of the values previously obtained, for it allows us to know, owing to Townsend's results, the charge on an atom in electrolysis, and from it to deduce immediately the constant of Avogadro, the number of molecules contained in a given volume of a gas. The results are well in agreement with the values of this constant (in general a little greater), which we can directly deduce from the kinetic theory of gases.

Here is an important group of concordant indications, all of absolutely distinct origin, which show without doubt the granular structure of electric charges, and consequently the atomic structure of matter itself. The measurements which I have just enumerated allow us to establish, in great security, the hypothesis of the existence of molecular masses.

I seek to point out here this extremely remarkable result, which belongs without doubt to some fundamental property of the ether and of the electrons, that all these electrified centres, whatever may be their origin, are now identical from the point of view of the charge which they carry.

It is necessary for us to penetrate further into their properties, into their relations with material atoms, to determine their relative sizes, in order to add among others to the more exact ideas which we possess in this field, that the electrons, or negative cathode corpuscles, are all identical not only from the point of view of their charge, but also from the point of view of their dynamic properties and of their masses. We are unhappily not so well informed in regard to the positive centres.

III. *Inertia and Radiation*

(11) *The Electromagnetic Wake.*[1] Before going farther it is important to point out what we can draw from the point of view to which we have now come. Electrified centres, whose existence is experimentally proven, whose charge we know in absolute units, are movable with respect to a fixed ether defined according to the equations of Hertz, without its having been necessary for us to have recourse to dynamic principles to arrive at this point of view.

To what extent can the known properties of matter be deduced from these two ideas of the electron and the ether, and is it necessary to add something to them in order to build up a synthesis? We are going to see rapidly and definitely from our idea of the electron, how it is sufficient to represent at the same time the inertia of matter, its dynamic properties, also how it can emit and absorb the radiations which the ether transmits.

The possibility of conceiving of inertia, mass, not as a fundamental idea, but as a consequence of the laws of electromagnetism, is a conception which owes its origin to an important memoir published in 1881 by Professor J. J. Thomson.[2] He studies there, basing his assumptions on the existence of the displacement currents of Maxwell, the electromagnetic field accompanying an electrified sphere in motion. This motion implies a change in the electric field at a point fixed with respect to the medium, and this displacement current immediately produces a magnetic field according to the ideas of Maxwell. The necessity of a convection current is pointed out later. The magnetic field thus produced, identical with that of an element of current parallel to the velocity of the moving charge, is proportional at each point to that velocity, at least, if it does not approach too nearly to that of light.

The creation of a magnetic field at the time of setting the charged centre in motion implies an expenditure of energy, energy of self-induction of the convection current, proportional to a first approximation to the square of the velocity, for those velocities which are small compared to the velocity of light. It is thus an expression of the same form as that of ordinary kinetic energy. A part, at least, of the inertia of an electrified body, of its capacity for kinetic energy, is thus a consequence of its electric charge.

Moreover, the magnetic field thus produced, and the electric field as well, modified by the velocity as it approaches more nearly to that of light, constitute around the electrified centre in translation a wake which accompanies it in its translation through the ether without change so long as the velocity remains constant. It is besides neces-

[1] *Le Sillage Electro-magnétique.*
[2] J. J. Thomson, *Phil. Mag.* t. 11, p. 229. 1881.

sary that an external action should intervene in order to modify the energy of this wake and consequently to increase or diminish the velocity. This implies, in the absence of all other kinetic energy than this of electromagnetic origin, corresponding to the production of the wake, by the law of Galileo on the conservation of the velocity acquired, in the absence of action of all external fields of force, that an electrified centre possesses inertia by the fact alone that it is electrified.

It is the immovable ether, the electromagnetic medium, which serves as a fixed support for the axes with respect to which the principle of inertia is applicable, and of which the ordinary mechanics limits itself in affirming the existence by saying: there exists a system of axes, determined by a nearly uniform translation with respect to which the principle of Galileo is exactly verified.

(12) *The Absolute Motion.* If we are able, from the actual point of view, to conceive of the ether as supporting these Galilean axes, it does not necessarily follow that the electromagnetic phenomena enable us to arrive at this absolute motion. It seems, on the contrary, so far, that static experiments, carried on in a material system by an observer carried along with it with a uniform motion of translation, do not allow, whatever may be the degree of accuracy of observation, the detection of a relative motion of the ether with respect to matter.

Larmor, and more completely Lorentz, have shown that there exist in the system actions of electromagnetic origin; it is possible to establish in a complete manner a static correspondence (relating to the positions of equilibrium or to the black fringes in optics) between the system in motion and a system fixed with respect to the ether, by means of a change of variables which preserves for the equations of the medium for a moving system the exact form which they possess for a system at rest.

The two systems differ from one another in that the moving system is slightly contracted compared with the fixed system in the direction of the resultant motion by an amount always very small, proportional to the square of the ratio of the velocity of motion to the velocity of light. This contraction affects equally all the elements of the moving system, *i. e.* the electrons themselves, if we admit with Lorentz that the interior actions of these electrons are solely electromagnetic actions or are modified in the same manner by the translation, — with the result that observation cannot prove this contraction any more than it can prove the general dragging of the ether. These elements behave as though they belonged to a corresponding fixed system. Thus is found an explanation of the negative results of experiments undertaken to show the absolute motion of the earth, by Michelson and Morley, Lord Rayleigh, Brace, Trouton, and Noble, if one admits that all the internal forces of matter are of electromagnetic origin, and that the energy is entirely divided between the two fields, electric and magnetic.

We shall see, however, farther on that it is difficult to eliminate in this way all other forms of energy, all other forces, such as gravitation; and it would then be necessary to admit with Lorentz, in order that the correspondence between the two systems should actually subsist, that in the moved system the forces and masses of different origins are modified exactly as the electromagnetic forces and masses, an hypothesis too complicated and arbitrary in the actual state of the question.

But this does not seem to be a necessary consequence: it appears probable that these actions, foreign to electromagnetism, and necessary at the interior of the electron in order to give stability and in order to represent gravitation, and which are probably connected with one another, do not intervene in a sensible manner in the negative experiments referred to above, and that everything transpires as if the electromagnetic forces alone played a rôle, alone existed.

We shall see farther on that perhaps experiments of another kind than those referred to here, for example, some dynamic measurements bringing in a relative motion of the system moved, or some static experiments bringing in gravitation, would enable us to understand the absolute motion, the axes bound to the ether, instead of conceiving simply of their existence.

(13) *Electromagnetic Inertia.* The problem of the electromagnetic wake accompanying an electrified sphere or ellipsoid in the ether has been taken up since J. J. Thomson by Heaviside and Searle.

Max Abraham has shown their results to consist approximately of a numerical factor when, instead of supposing the body to be a conductor having a surface charge, we suppose its charge to have a uniform volume distribution.

Among the more important results contained in this solution of J. J. Thomson's problem, I will point out these: that in the case of a conducting sphere, the charge remains uniformly distributed on the surface whatever may be the velocity, and that in all cases the electric field at a distance tends to become more and more concentrated in the equatorial plane with respect to the direction of the velocity in proportion as this velocity approaches that of light.

Moreover the kinetic energy which it is necessary to expend at the moment of putting it in motion in order to create the electromagnetic wake ceases to be proportional to the square of the velocity, and increases indefinitely as the velocity approaches the velocity of light-waves; the law of the increase of this kinetic energy with the velocity, the energy of self-induction of the current to which the charged body in motion is equal, may be easily deduced by Searle's solution.

Without any other hypothesis than that of its electric charge, the electron is found to have inertia defined as capacity for kinetic energy, but with a particular law of variation of this as a function of the velocity, and this inertia appears to approach infinity as the velocity approaches that of light.

The behavior of this law depends very little on the hypothesis made as to the form of the electron and the distribution of the electric charge which it carries. In all cases it is found to be impossible to give the electron a velocity equal to that of light, at least permanently.

Instead of considering with Max Abraham the electron to be spherical at all velocities, Lorentz admits it to be spherical when at rest and to have a uniform distribution of charge; but if all internal forces are solely electromagnetic or act as such, we have the view that the electron is flattened in the direction of motion by a quantity proportional to the square of the ratio $\left(\beta = \dfrac{v}{V} \right)$ of its velocity to that of light, becoming an ellipsoid of revolution, the equatorial diameter remaining equal to that of the original. This leads, as we shall see, to a law of inertia different from that of an invariable sphere.

We shall likewise see that it does not appear to be necessary to assign to the electrons, the negative ones at least, any other inertia than this in order to account for the dynamic properties of the cathode rays; however, experiments are not yet sufficiently exact to allow us to infer the form of the electron itself, which depends on the law of the variation of the kinetic energy with the velocity.

(14) *Two Problems.* We have examined, so far, only the case of an electron in uniform motion in the absence of any external electromagnetic field capable of modifying the motion of the electron by giving it an acceleration.

The general problem of the connection between the ether and the electron, which probably represents the most important of the connections between ether and matter, is double.

In the first place, what is the electromagnetic disturbance in the ether accompanying any given motion of the electrons whatsoever?

In the second place, what motions would free electrons have if displaced in an external magnetic field superimposed on that which constitutes their wake?

(15) *The Velocity Wave — The Acceleration Wave.* We actually possess all the elements necessary for the solution of the first problem, in which the motion is uniform in a particular case. Lorentz has given in a very simple form the general solution by the use of a delayed potential.

Each element of the charge in motion is determined by its position, its velocity, and its acceleration at the time T, the electric and magnetic fields at the time $T+t$, on a sphere having for its centre the position at the time T and for radius the path passed over by light during the time t.

Lorentz has given in this way the expressions for the two electric and vector potentials from which the fields can be deduced by the well-known formula. The complete expressions for these fields have been given for the first time, I believe, by Lenard ; I obtained them independently at the same time as Schwartzschild by putting them in the following form.

The expressions for the two fields consist of two parts: the first depends solely on the velocity of the element at the time T and contributes to form the wake (*sillage*) which accompanies the electron in its motion; I shall call this the *velocity wave*. This velocity wave, which exists only in the case of uniform motion, has its electric field always directed toward the position which the element of charge will occupy at the time $T+t$, if it had retained from the time T the velocity which it had at that moment. Schwartzschild calls this position the point of aberration. It coincides with the true position of the moving element at time T if the motion has been uniform. The other part of the two fields is proportional to the acceleration projected on the direction of propagation, and the directions of the two fields are there perpendicular to one another, and perpendicular to the radius, at the same time the two electric and magnetic fields represent equal energies per unit volume; they have all the characteristics of a *radiation* which is freely propagated in the ether. I shall call this part the *acceleration wave*. Moreover, the intensities of the fields in this case vary inversely as the distance from the centre of emission, the energy represented by this wave does not tend toward zero as the time T increases indefinitely ; there is thus energy radiated to infinity by the acceleration wave.

The velocity wave, on the contrary, in which the fields vary inversely as the square of the radius Vt, does not carry any energy to infinity: the energy of the velocity wave accompanies the electron in its motion and corresponds to its kinetic energy.

(16) *Radiation implies Acceleration.* We can conclude from this that when an electrified centre experiences an acceleration, and only then, it radiates to infinity in the form of a transverse wave, electromagnetic radiation, a definite quantity of energy, proportional per unit of time to the square of the acceleration.

The origin of electromagnetic radiation, of all radiation, is, then, in the electron undergoing acceleration. It is through the electron that matter acts as the source of Hertzian or light waves. All acceleration, all change which takes place in the state of motion of electrons, result in the emission of waves. The character of the emitted waves changes naturally according as the acceleration is abrupt, discontinuous, or periodic.

In the first case, realized, for example, in the sudden stopping of the negative electrons, or corpuscles, by the anti-cathode, the radiation consists of an abrupt pulse whose thickness is equal to the product of the velocity of light into the time taken to stop them, and which gives us a good representation of the Roentgen rays or of the rays from radioactive substances.

If the acceleration is periodic, on the contrary, as in the case when the electron revolves around an electrified centre of opposite sign to itself, the acceleration is periodic, and the radiation emitted constitutes a light-wave whose length is determined by the period of revolution of the electron.

The solution of the first of the two fundamental problems thus appears complete and raises no difficulty.

IV. *Dynamics of the Electron*

(17) *Maxwell's Idea.* The inverse problem is less simple. It consists in finding the motion, the acceleration which a movable electron experiences in electric or magnetic fields of given intensities; it is, properly so to speak, the problem of the dynamics of the electron.

The equations which solve this problem ought to consist, like the equations of ordinary dynamics, of two kinds of terms: one of these dependent on the external fields, which produce their actions on the electron, and are analogous to the external forces in dynamics; the other, representing forces dependent on the motion itself, and producing a resistance to motion, similar to the forces of inertia.

The terms corresponding to external actions, the forces, have been obtained by Lorentz following a method which was the natural continuation of Maxwell's idea as to the possibility of a mechanical explanation, otherwise indeterminate, by the facts of electromagnetism. The analogy to the equations of electrodynamic induction, and to the equations of Lagrange, appeared to justify such an explanation, and it was natural to continue to look upon the ether-electron system as a mechanical system, and to apply to the motions of electrified centres Lagrange's equations, deducing thus the forces exerted on the electrons by its electric and magnetic energies considered as corresponding to the potential and kinetic energies of a mechanical system, substituted in the ether. We are thus led to apply to the medium, ether, in consideration of the fundamental notions of force and mass, which they imply, the equations of material dynamics, deduced from principles founded on observations of matter only, always taken in mass and without an appreciable amount of radiation.

(18) *Ether in Matter.* We extend thus, by a bold deduction, these principles to a region for which they have not been designed, and thus admit implicitly the possibility of a material representation of the ether. However, as I have already pointed out, an attempt at such a representation raises many difficulties, and the efforts so far made to extend these principles in a more precise manner have not been successful. The most profound attempt, that of Lord Kelvin, the gyrostatic ether, lends itself rigorously only to the representation of the propagation of periodic disturbances in the ether, but makes impossible the existence of a permanent deformation, necessary, however, for the representation of a constant electrostatic field. The gyrostats would turn back again at the end of a finite time, and the system would cease to react against a deformation which has been imposed. Moreover, it would appear impossible to include in this conception the permanent existence of electrons, centres of deformation in the medium.

To get around this difficulty, Larmor had occasion, in the material image which he proposed for the ether, to superimpose on the gyrostatic system of Lord Kelvin the properties of a perfect fluid, of which the displacements representing the magnetic field should be at each instant irrotational in order not to produce an electric field by the rotation of the gyrostats present in the medium. But a great difficulty is added to the preceding: if the motion of a fluid satisfies at every moment the condition of being irrotational for infinitely small displacements, it is not so for finite displacements, and a magnetic field could not continue to exist without giving rise to an electric field.

I believe it impossible to overcome these difficulties and to give a material image of the ether, whose properties are entirely distinct, and probably much more simple than those of matter.

(19) *Action and Reaction.* Let us, however, retain this view in order that we may meet new difficulties. By means of Lagrange's equations Lorentz obtains two external forces acting on each electron in motion, two terms representing the action of the electromagnetic field.

One force is parallel to the electrostatic field; it is the ordinary electric force, due to the superposition of the electric field produced by the electron on the external electric field: the other is perpendicular to the direction of the velocity of the electron and of the external magnetic field; it is the electromagnetic force analogous to the force of Laplace exerted by a magnetic field on an element of current, and due to the superposition on the external magnetic field of the magnetic field produced by the electron during its motion. This double result includes all the elementary laws of electromagnetism and of electrodynamics, if we consider the current in ordinary conductors as due to the displacement of electrified particles.

We easily see that the forces thus obtained, exerted on the electrons by the ether, *i. e.* on the matter which contains them, do not satisfy the principle of the equality of action and reaction, if we consider all the forces which act at the same moment on all the electrons constituting matter. In the case of a body which radiates in an unsymmetrical manner, a recoil, an acceleration, is produced which is not compensated at the same moment by an acceleration set up in another portion of the matter. Later, at the time that the emitted radiation meets an obstacle, the compensation is made (but only in a partial manner if all the radiation is not absorbed) by means of the pressure which the radiation exerts on the body which receives it; a pressure whose existence is shown by experiment.

The equality of action and reaction has never been verified in similar cases, and it adds no difficulty to this subject if we do not seek to extend the principle beyond the facts which suggested it.

(20) *Quantity of Electromagnetic Motion.* If we could nevertheless realize this extension of the principle, an extension somewhat arbitrary, we should be led not only to apply this principle to matter, but to suppose the ether to have a quantity of motion which would be that of a material system to which we compare it.

Poincaré has shown that this quantity of electromagnetic motion ought to be, at every point in the ether, in direction and in magnitude, proportional to Poynting's vector, which gives at the same time a definition of the energy transmitted through the medium.

By starting with this idea of the quantity of electromagnetic motion, Max Abraham has been able to calculate the terms, put to one side by Lorentz, which depend on the motion of the electron itself, its force of inertia, by the variation of the quantity of electromagnetic motion contained in its train. He was led for the first time, by the form of the terms which represent this force of inertia, to the notion of an unsymmetrical mass as a function of the velocity.

(21) *Quasi-Stationary Motion.* The calculation can be completely made only in the case, always realizable from the experimental point of view, where the acceleration of the electron is so small that its train can be considered at each instant as identical with that of an electron having the actual velocity, but whose motion has been uniform for a long time. This is what Abraham calls a quasi-stationary motion. In this case, the train is entirely determined at each moment by the actual velocity of the electron, also the quantity of electromagnetic motion which it contains, and consequently the variation of this quantity which represents the force of inertia. The condition of quasi-stationary motion is simply that in the neighborhood of the electron, where the quantity of electromagnetic motion is localized, the wave of acceleration may be neglected in comparison with the velocity wave.

(22) *Longitudinal Mass and Transverse Mass.* We find under these conditions that the force of inertia is proportional to the acceleration with a coefficient of proportionality analogous to mass, but which is here a function of the velocity, and increases indefinitely, like the kinetic energy, as the velocity tends to approach that of light. Moreover, this electromagnetic mass differs for the same velocity, according as the acceleration is parallel or perpendicular to the direction of the velocity. There is, corresponding to the direction, a longitudinal and a transverse mass. Mass is then no longer a scalar quantity, but has the symmetry of a tensor parallel to the velocity. No experimental fact yet allows us to verify this dissymmetry of the mass of the electrons, which becomes evident only when the velocity is of the same order as that of light, but the variation of the transverse mass with the velocity has been proven by Kaufmann for the β rays of radium, which consist of particles identical with the cathode rays. It is sufficient to compare the deviations of these rays in the electric and magnetic fields perpendicular to their direction in order to deduce, by application of the equations of the dynamics of the electron, their velocity and the ratio of the charge to the transverse mass of the particles which compose them. This ratio decreases as the velocity increases, and, if we consider as fundamental the principle of the conservation of electricity, we conclude from it an actual increase of the transverse mass according to a law easy to compare with that which the theory gives for the electromagnetic mass.

(23) *Matter of the Philosophers.* But, before discussing the result of this comparison, I wish to point out a logical difficulty raised by the course which we have followed: we are accustomed to consider as fundamental the ideas of mass and force, built up in order to represent the laws of motion of matter; we, *a priori*, conceive of mass as a perfectly invariable scalar quantity.

Now, let us suppose the possibility of a material representation of the ether: we apply to it the equations of material dynamics, and we are led to admit for the electrons, which form a part of matter, and consequently for matter itself, a dissymmetrical mass, tensorial and variable.

To what, then, should the equations of ordinary dynamics apply, and what are the ideas considered as fundamental which they imply? To an abstract matter, the matter of the philosophers, which could not be ordinary matter, since it is inseparable from electric charges, and which is probably made up of an agglomeration of electrons in periodic motion, stable under their mutual actions? Or to the ether? But we have no idea of what can be its mass or motion.

It is, indeed, rather the ether which it is necessary to consider as fundamental, and it is then natural to define it initially by those properties of it which we know, that is to say, by the electric and magnetic fields, which it is possible to arrive at, as I have already remarked, without admitting at any time the laws of dynamics, the ideas of mass and force under their ordinary form. We will find this last to be a derived and secondary idea.

V. *Electromagnetic Dynamics*

(24) *Change of Point of View.* It seems thus much more natural to reverse the conception of Maxwell and to consider the analogy which he has pointed out between the equations of electromagnetism and those of dynamics under Lagrange's form as justifying much more the possibility of an electromagnetic representation of the principles and ideas of ordinary, material mechanics, than the inverse possibility.

It is necessary then for us to solve our second problem, that of the dynamics of the electron, of its motion in a given external field, without having recourse to the principles of mechanics, by purely electromagnetic considerations.

Hertz's equations, which permit a solution of the first problem, are here not sufficient, and we have need of a more general principle, which assumes not the motion of the electrons given, but that determines it.

(25) *The Law of Stationary Energy.* We will use this principle under a form indicated by Larmor, and which we can look upon as a generalization of the known laws of electrostatics and of electrodynamics. We know that the distribution of electric charges and electric fields in a system of electrified bodies is always such that the electrostatic energy W_e, contained in the medium modified by the field, is a minimum. The analogous principle holds for the magnetic field produced by currents of given intensities. The energy W_m localized in the magnetic field is less for the real distribution of it than for all other distributions satisfying the condition that the integral around a closed line is equal to 4π times the intensities of the currents inclosed by the line.

If displacements are possible, the conductors maintained at constant potential are in stable equilibrium if the electrostatic energy is a maximum, and the currents of given intensities are likewise in stable equilibrium if the energy of their magnetic field is a maximum. In all cases of maxima and minima, an infinitely small modification of the system from the configuration of equilibrium produces a zero variation in the energy: it is stationary.

(26) *General Principle.* When, instead of remaining permanent, the state of the system is variable, and if there are represented necessarily at the same time the two kinds of fields, we seek to find how,

as in the permanent case, an expression which remains stationary, that is to say, the variation of which is zero when supposed slightly modified, can start from its real state. We are thus led to replace the energies W_e, W_m, which play this rôle in the permanent case, by an integral taken with respect to the time, and which represents not the sum of the energies, since this quantity, equal to the total energy, ought to remain constant if only electromagnetic action come in, but their difference:

$$\int_{t_0}^{t_1} (W_e - W_m)dt,$$

an integral which remains stationary for all virtual modifications of the system, such modifications being subject to the condition of disappearing at the limits t_0 and t_1 of the integral, exactly as in the analogous principle of Hamilton in mechanics. The principle of zero variation just announced, and which we will consider as the result of an induction based entirely on electromagnetic principles, allows us in fact to find three of Hertz's equations, if we admit the three others as an imposed interconnection of the system, and furnishes in the most simple manner the solution which we have obtained for the first problem by means of these equations. Moreover, the motion of the electrons supposed given only at the times t_0 t_1 comes into the integral, and the condition that this must be stationary allows us to find the law of the motion during the interval, by starting from a principle whose signification is purely electromagnetic. We obtain thus exactly the results of Max Abraham; the equations of motion contain terms which depend first on the motion of the electron, and are proportional, in the hypothesis of quasi-stationary motion, to its acceleration, having coefficients that are functions of the velocity which we will call the longitudinal and transverse masses of the electron; also some terms depending on the charge, and on the external fields, which we will call the forces, and we find that they coincide with those given by Lorentz. The external motion of the electron is thus determined by the actual electromagnetic state of the system.

(27) *The Process in the Electron.* In order to simplify the analysis and to avoid considering the motion of rotation of the electron, I will consider it as a cavity in the ether; the volume integrals which express the energies W_e, W_m of the electric and magnetic fields extend only over the space external to the surface which bounds the cavity. We can suppose as a special condition outside of the electric charge that the form of this surface is fixed, spherical for example, due to an unknown action of nature, and we find the equations of Abraham for the longitudinal and transverse masses of a spherical electron.

But we can suppose a more simple condition, implying only a fixed volume of the cavity on account of the incompressibility of the ex-

ternal ether; if we seek, then, what is, in the case of uniform translation, the form that the electron would spontaneously take in order
to satisfy the condition of zero variation, we find precisely the oblate
ellipsoidal form assumed by Lorentz, with this difference, that the
equatorial diameter increases with the velocity instead of remaining
constant, as Lorentz considers it; this constancy implies a diminution
of the volume as the velocity increases. The equations which express
in this case the variation of the longitudinal and transverse mass
with the velocity are different from those of Abraham and Lorentz,
although giving always an indefinite increase of the two masses as
the velocity approaches that of light.

The equations thus obtained for the ratio $\dfrac{m}{m_0}$ of the transverse mass

m, the only one so far accessible to experiment, to the mass m_0 for

very small velocities, as a function of the ratio $\beta = \dfrac{v}{V}$ of the velocity

of the electron to that of light are:
(1) Invariable spherical electron,

$$\frac{m}{m_0} = \frac{3}{4} \ \psi \ (\beta) = \frac{3}{2\,\beta^2} \left[\frac{1+\beta^2}{2\beta} L\frac{1+\beta}{1-\beta} - 1 \right]$$

(2) Variable Electron $\begin{cases} \text{Equatorial diameter constant } \dfrac{m}{m_0} = (1 - \beta^2)^{-\frac{1}{2}} \\[2ex] \text{Volume constant} \qquad\qquad \dfrac{m}{m_0} = (1 - \beta^2)^{-\frac{1}{3}} \end{cases}$

(28) *Comparison.* The researches of Kaufmann are not yet exact
enough to determine which of these equations represents most nearly
the experimental variation of the ratio $\frac{e}{m}$ with the velocity. In
order to make the comparison, I have used a process similar to that
of Kaufmann, who eliminated the two electric and magnetic fields used
to deviate the β rays, seeking to obtain the best concordance possible between the experimental variation of $\frac{e}{m}$ and the theoretical
variation calculated on the hypothesis that the mass is entirely
electromagnetic.

In order to make this elimination, I draw the two experimental and
theoretical curves representing $\frac{e}{m}$ as a function of β, on logarithmic
coördinates, and seek for what relative positions of the curves we obtain the best correspondence. The results are given for the three
theoretical equations and the same series of experimental values.
The experimental points corresponding to four different series are
given by Kaufmann, and we see that they correspond equally well
with the three theoretical curves.

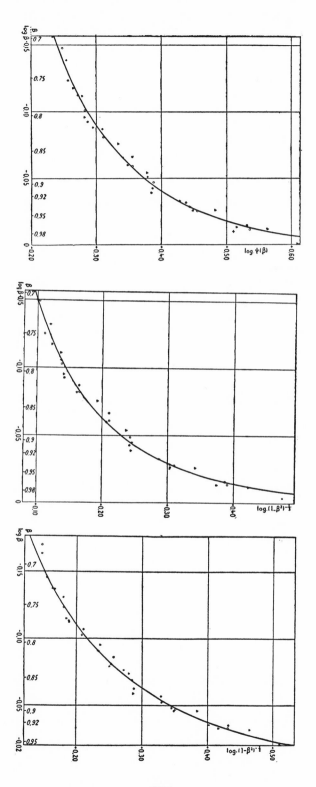

The more important values from the point of view of choice of equations are those corresponding to values of the velocity very near to that of light, and which amounted to ninety-five per cent of it in Kaufmann's experiments. But the β rays are then very little deviated, and exact measurements are extremely difficult.

It would be extremely important to determine the longitudinal mass by the use of an intense electric field parallel to the velocity of the electron, furnishing to it a known energy and producing a variation of the velocity, which if measured would give the longitudinal mass.

(29) *Matter and Electrons.* But if the accuracy of experiment is not sufficient to determine completely the law, the agreement with the equations, obtained by supposing the mass to be entirely electromagnetic, is so good that we can reasonably conclude that cathode particles constituting the β rays have no mass other than that due to their electric charges or the train which they carry with them in their motion through the ether.

It is interesting to extend the same result to ordinary matter by conceiving it as made up of an aggregation of electrons of both signs; it is unreasonable on the other hand to apply to two phenomena so nearly identical as inertia of ordinary matter and that of the cathode particles, two entirely distinct explanations, of which the one, the electromagnetic explanation, is definite and confirmed by experiment, while the other remains entirely unknown.

The inertia of a similar aggregation of electrons should be equal to the sum of the partial inertias because of the great distance of the electrified centres from one another compared to their radii, which one can calculate by supposing all their inertia electromagnetic.

In these conditions, the trains of the different electrons do not interfere appreciably, and we find thus the law of the conservation of inertia as a consequence of the conservation of the electrons in the transformations to which matter is subject. But the theory is not incompatible, on account of the interference of trains, with a slight disagreement between the inertia of an assemblage and the sum of the partial inertias.

The complexity of the atomic system to which we are led, each atom of the molecule containing probably a very great number of electrons, seems also to be a necessary consequence of the complexity of the luminous spectrum sent out from the atoms, by the electrons which they contain, when an external disturbance displaces the system from its state of stable periodic motion. In such a state the radiations emitted by the various electrons on account of the acceleration which keeps them in their intermolecular orbits compensate one another almost completely from the point of view of energy radiated; so that there is in general no decay of the periodic intermolecular motion.

This conception, this electronic theory of matter in which matter becomes, at least partially, synonymous with electricity in motion, appears to account for an enormous number of facts, which increase constantly under the efforts of physicists impatient to contemplate in a less primitive form the synthesis which it promises to bring forward.

(30) *Stability of the Electron.* The fundamental conception, that of the electron, does not go without raising difficulties still further, besides the impossibility already pointed out of representing to ourselves by material images its displacement with respect to the ether. It seems necessary to admit something else in its structure than its electric charge, an action which maintains the unity of the electron and prevents its charge from being dissipated by the mutual repulsions of the elements which constitute it. The form of the electron is determined by some relation which insures its stability, the condition of incompressibility of the medium being insufficient, since the spherical form corresponds only to unstable equilibrium for an electrified body of given volume in which no force opposes the deformation.

This condition, which belongs to some fundamental property of the medium, determining the charge carried by the electrons, all identical from this point of view, is perhaps closely connected with the third mode of activity of the ether, a third form of energy, the gravitational form, of which our principle of stationary energy ought to take account by the addition of terms to those expressing the electrostatic energy, but of infinitely smaller magnitude.

(31) *Gravitation.* Gravitation remains obstinately outside of our electromagnetic synthesis; the Newtonian forces not only do not appear to be propagated with the velocity of light, but also it seems difficult to found them on electromagnetism without modifying profoundly our fundamental ideas in regard to field and quantity of electricity and the possibility of an attraction of one aggregation of neutral electrons for another aggregation of the same nature.

It appears probable that gravitation results from a mode of activity of the ether and a property of electrons entirely different from the electromagnetic mode, and we must admit besides electric and magnetic energies, a third distinct form, that of gravitation.

It remains to understand how it is possible, and what is the significance of the equivalence, the passage of this third form into one of the first two. Also we are no more capable of understanding, outside of the formal equations which express it, the connection between the electric and magnetic energies themselves and their transformations, the one into the other, by means of the electrons.

(32) *An Experiment Necessary.* It does not seem impossible to connect the forces of cohesion with electromagnetism, especially

from the point of view of the mutual attractions which orientation causes in the constitution of crystalline media, on account of the complex electric and magnetic fields which surround a system of electrons in its immediate vicinity.

Gravitational forces alone remain distinct, superimposed on the electromagnetic forces, and no difficulty comes from this on account of the negative results of the experiments undertaken to show the absolute motion of the earth.

The negative results can be explained, as we shall see, if all the internal forces of matter are of electromagnetic origin; but gravitational force, alone different, can be superimposed on them without introducing an appreciable modification of this result, for its intensity is extraordinarily small compared to electromagnetic actions, even if there is no mutual compensation between them, and in all the experiments in question, interference of light or equilibrium of an elastic system, the gravitational forces play no appreciable rôle.

It would be, indeed, important to obtain a condition in a case of equilibrium where the forces of gravity would play an important part, and if the equilibrium remains independent of the total motion to nearly the second order, if we could only observe the mutual motion to this order of precision, it would be necessary to conclude that the forces of gravitation also are modified by motion of translation in the same manner as the electromagnetic forces, since the equilibrium between the two kinds of forces is not disturbed, and this would be an important indication of the necessity of an electromagnetic representation of gravitation. We would be able, for example, if the sensibility allowed it, to perform the experiment of Trouton and Noble by suspending the condenser with a bifilar to the pan of a balance instead of by an elastic fibre.

Since this test has not been made, since experiments designed to show the absolute motion have not involved weight, it would be more reasonable to consider gravitation as a force distinct from electromagnetic action, which acts at the interior of the electrons in order to insure their stability, without its being possible actually to imagine in what manner we can seek a more profound knowledge of the ether and of the electrons which it incloses.

It does not seem, in any manner and for many reasons, that this can be of the nature of a material and mechanical representation of the ether.

VI. *Cathode Rays*

(33) *The Ratio e/m.* Before examining the consequences involved in the electronic conception of matter, I should like to examine a few points relative to the electrons of two kinds. Those which we know the better, the more intimately, are the negative electrons, which

are always identical with one another in all their properties, whatever may be the matter which has furnished them. We have already seen how the direct measurement of the charge leads always to the same result, The mass, both the longitudinal and transverse mass, having the same value for small velocities, can be determined by the measurement of the ratio of the charge to the mass.

The results obtained for this ratio in the case of cathode rays show some quite marked divergences when different methods of measurement are employed. The first values were given by J. J. Thomson by combining the magnetic deviation of the rays with a measurement of the energy which they possess by means of the heat produced in a thermoelectric couple which receives them, or by combining this magnetic deviation with the deviation in an electrostatic field. The ratio $\frac{e}{m}$ furnished by this second method, the more accurate of the two, is approximately 10^7 electromagnetic units C.G.S.

Another method first pointed out by Schuster was used successively by Kaufmann and Simon. It consists in combining the magnetic deviation with the measurement of the difference of potential under which the rays are produced, considering that this difference of potential is that which exists between the cathode and anode. This hypothesis admitted, the method is capable of great accuracy, and the results which it gives appear to agree with the limiting values, for small velocities of the ratio $\frac{e}{m}$ for the β rays, although the method employed by Kaufmann in this last measurement is different from that of Schuster. The number obtained by Simon is 1.865×10^7, nearly double that of J. J. Thomson. The explanation proposed by the latter for this disagreement, according to which the cathode rays are not produced by the total difference of potential between the cathode and the anode, but originate in a region situated in front of the cathode, does not, however, appear satisfactory, since it does not account for the constancy of the results of Kaufmann and Simon when the conditions of the experiment, the difference of potential in particular, were varied between large limits.

A means of deciding the question would consist in performing a type of experiment already used by Lenard, by subjecting the cathode rays, after their production, to a supplementary and known fall of potential, and determining by the modification which would result in their magnetic deviation the initial fall of potential under which they had been produced.

(34) *The Cathode Corpuscle.* However it may be, we can, owing to the results of Kaufmann, affirm the identity of the cathode rays already found independent of the gas and the electrode contained in the Crookes tube, with the β rays of radium. The measurements

by J. J. Thomson and Lenard of the negative charges emitted by a negatively charged metallic surface under the action of light and of those spontaneously emitted by incandescent bodies also show an identity with the cathode rays. Wehnelt has recently shown that the oxides of the alkaline earths possess in an extraordinary degree this property of spontaneously emitting cathode rays at high temperatures, and furnishes a means of performing, on this particular kind of rays, simple and exact measurements.

Finally, we know that the magnitude of the Zeeman effect, in the case where the spectrum lines considered present the appearance of a normal triplet, leads to the conclusion that the light corresponding to these lines is emitted by negatively electrified centres, present in matter and having the same ratio $\frac{e}{m}$ as the cathode rays.

Moreover, the magnitude of this ratio, one thousand to two thousand times greater than for the hydrogen atom in electrolysis, leads us, as a consequence of the identity of charges established by Townsend, to consider the mass of the cathode corpuscle as one thousand times smaller at least than an atom of hydrogen; a result in perfect agreement with the conception which makes material atoms an agglomeration of electrons of two kinds. On the hypothesis that the mass is entirely of electromagnetic origin, the knowledge of the ratio $\frac{e}{m}$ gives for the electron a sufficiently small radius (10^{-13} centimeters about) in order to be, conformably to our conception also, negligible in comparison with atomic dimensions.

(35) *Flames.* The small mass of the cathode corpuscle, and the possibility of separating from matter electrified centres a thousand times smaller than the smallest atom, is confirmed by the mobility of the negative ions in flames. We obtain enormous mobility compared to that observed in gases at ordinary temperatures, and the methods of the kinetic theory of gases permits us to calculate, by means of this experimental mobility, that the movable negative centres in flames have a mass about a thousand times smaller than the hydrogen atom, and should consequently be identical with the cathode corpuscles. At ordinary temperatures the negative ions are less mobile because the cathode corpuscles surround themselves with neutral molecules by simple electrostatic attraction, and form an agglomeration which the feeble agitation allows to remain stable.

VII. *Positive Electrons — a Rays*

(36) *Goldstein Rays. a Rays.* Our knowledge of the structure of positive charges is much less advanced than for the negative. Two important cases show us the existence of positively charged particles, besides the positive ions in conducting gases, which at ordinary temperatures consist of an agglomeration of neutral molecules around

a charged centre: these are the Kanalstrahlen of Goldstein, an efflux of positive charges toward the cathode, the electric and magnetic deviations of which lead to values for the ratio of $\frac{e}{m}$ varying between wide limits, but always several thousand times smaller than for the cathode rays. The mass of these positive centres is of the order of that of the atoms. The a rays of radioactive bodies, easily absorbed, and particularly easy to observe in the case of polonium and the active bismuth of Marckwald, appear to be, in fact, Kanalstrahlen. The mass of the positively charged particles which constitute these rays is of the same order as that of the hydrogen atom, and their velocity does not exceed 20,000 to 25,000 kilometers per second, so that it is impossible to verify whether their mass is entirely electromagnetic or not. Can we consider them as electrons as simple as the negative corpuscle itself, or are they of much more complex structure; are they, for example, atoms or molecules which have lost a cathode corpuscle?

(37) *Electrons or Atoms.* On the first hypothesis, the great mass of the positive centres would lead us to assign them dimensions much smaller than the cathode corpuscles themselves, the electromagnetic mass of an electrified sphere being inversely proportional to its radius. One is thus led to the result that an electron possesses inertia, I will not say weight, inversely proportional to its radius. H. A. Wilson thinks to find an argument in favor of this conception of a very small and consequently very inert positive electron in the observation that the a rays are much less easily absorbed than the β rays of the same velocity.

Many other reasons lead us to adopt the contrary hypothesis that an a particle is very complex and little different from an atom. Rutherford has given serious reasons for identifying the a particle with the helium atom deprived of a cathode corpuscle; also Stark gives experimental reasons referring to the emission spectra of positive centres in vacuum tubes, which imply a complex structure. Finally the theory of the disruptive discharge attributes the production of cathode rays in part at least to the impact against the cathode of particles which constitute the Goldstein rays; an electron smaller than the cathode particle itself seems scarcely able to produce a surface disturbance sufficiently intense, while on the other hand, an atom, unable to penetrate another atomic structure, and projected with a high velocity, would produce by its impact a considerable local disturbance.

(38) *The Positive charge of the a Rays.* It is perhaps by this considerable disturbance produced by the a or canal rays in matter which they meet that one can explain the interesting fact that the positive charge of the a rays has not been directly shown so far by the negative charge which a polonium salt should spontaneously

221

acquire if it emits only a rays. However high may be the vacuum around a piece of radioactive bismuth, or polonium, it does not acquire any charge, and loses rapidly, on the contrary, its positive or negative charge. Possibly one might explain this discharge by the ionizing action of the a rays on the gas, however rare. The passage of a particles, projectiles of large dimensions, through the surface of radioactive bodies from which they come, can play the same part as the impact of Kanalstrahlen on the surface of the cathode, and cause the emission of cathode rays of very little penetrating power, whose presence would suffice, added to that of the a rays, to prevent any permanent charge of the radioactive body, whatever may be its sign.

(39) *The Positive Electrons.* If the positive centres, as we know, ought not to be represented as free electrons, it seems, however, necessary to admit the presence of probable electrons which cause the neutralization of the negative charges in the atomic structure, but which for some reason come out of this structure with extreme difficulty, contrary to what is the case for the negative centres. Moreover, it would appear necessary in order that the theory of metals, which ascribes their conductivity to the presence of free electrified centres moving under the action of a field can take account of all the facts, the Hall effect in particular, of variable sign in different metals, that the centres of two kinds coexist in the metal, free to move about in all directions. These positive centres do not appear to be the metallic atoms themselves, necessarily immovable in order to maintain the solid framework of the metal. It is possible that the positive electron, which no known action in a gas can maintain separate from the atomic material, may be free in large numbers in the entirely different medium which constitutes the metal. Many problems present themselves here on the subject of the nature of the positive charges.

VIII. *Theory of Matter. Radioactivity*

(40) *Atomic Instability.* Let us examine now a little more closely the consequences to which we are led by the conception of matter as made up of electrons of two signs, of atoms formed of electrified bodies in motion under their mutual actions. From the first, — outside of gravitation, whose intensity is infinitely small compared to the electromagnetic forces in the interior of atoms which determine all the physical and chemical changes of state, — the elementary laws of action reduce to the forces of Lorentz, which allow us, as we have seen, to calculate the acceleration to which an electron is subjected as function of the electric and magnetic fields produced by the other electrons at the point where the first electron is situated. In the case

where the acceleration is sufficient for it to radiate an appreciable energy to a distance by means of the acceleration wave, it is probably necessary to bring in, by other terms in the equation of motion of the electron, some forces by which it can receive again the energy which it radiates, and which disappear in the case of quasi-stationary motion. It does not seem, however, in any experimental case that these corrective terms can become appreciable.

From the same point of view, the electrons in periodic motion in the material atom are necessarily subject throughout their closed orbits to accelerations which are accompanied by a radiation of energy borrowed from the internal electric and magnetic energy of the atom. This radiation must be extremely small, as in the simple case of several cathode corpuscles circulating at equal distances in the same orbit, and can be compensated for by energy obtained from external radiation. We can suppose that this continual radiation, much more important naturally when the atom, as the result of external shock, is displaced from its most stable equilibrium, is a cause of decay to the atomic structure and which at the end of a certain length of time ought necessarily to give the structure a fundamental rearrangement, as a top falls when its rotation has sufficiently diminished in velocity. A condition of instability is thus reached, the consecutive rearrangement being accompanied by a violent projection of certain electrified centres from the atom. This conception furnishes at least an image of radioactive phenomena, and the successive transformations in the life of an atom, an hypothesis of which has been advanced by Rutherford. It seems, however, that it is not necessary to admit a probable decay of atomic structures, sensible only for radioactive substances. The fact that the dispersion takes place as a function of the time according to a rigorous exponential law, the quantity which is destroyed in a given time being exactly proportional to the quantity present, seems to indicate that the substance not destroyed remains identical with itself. Perhaps the reorganization of the atomic structure might result from its accidental passage through a particularly unstable configuration, the probability that a like configuration should be reproduced being independent, in the mean, of the previous history of the atom, and the mean life of the latter would be short in proportion as this probability is great.

(41) *Internal Energy and Heat set Free.* A very simple calculation shows also that the stock of energy represented by the electric and magnetic fields surrounding the electrons contained in an atom is sufficiently great to supply for ten million years the evolution of heat discovered by Curie in the radium salts. As it appears now well established that the mean life of a radium atom is of the order of a thousand years, it results that the ten-thousandth part only of this reserve of energy is utilized during this especially active period in

the life of the atom. There is then no difficulty in conceiving how the enormous evolution of heat by radium can be ascribed to its internal energy.

No atom being free from this loss of energy due to the radiation of the electrons, one ought to expect on this hypothesis of decay a universality of radioactive phenomena, the atoms which we consider as actually stable suffering only an extraordinarily slow waste.

IX. *Electric Properties*

(42) *Polarization.* It remains now to show in a few words how the preceding conceptions lend themselves easily to a representation of the principal electric and magnetic properties of matter and make possible for the first time a theory of the disruptive discharge and of metallic conduction.

A common property of all forms of matter is electrostatic polarization arising from the variation of the specific inductive power with the nature of the substance.

This polarization results in a manner quite natural by the modification which an external electric field produces in the motions of the electron which constitute the atom. This modification is caused in the mean by an excess of positive centres on the side where the field tends to displace them and by an excess of negative centres on the opposite side. The system takes then on the average an electrostatic polarization.

(43) *Corpuscular Dissociations.* If the electric field becomes sufficiently intense, as, for example, during the passage of one of those brief pulsations which constitute the Roentgen rays, or during the passage through the atomic structure of an a or β particle of very great velocity, the modification produced may be very great, a cathode corpuscle may be separated from the structure which remains positively charged; there is produced thus a corpuscular dissociation which explains the conductivity acquired by insulating mediums under the action of Roentgen or Becquerel rays, and which manifests itself especially in gases, where the electrified centres thus freed can move more easily, although by electrostatic attraction on the neutral molecules, electrically polarizable, they surround themselves with a group of molecules which accompany them during their motion.

It seems well established that the negative ions in particular, also produced in a gas, have a cathode corpuscle for centre, since the penetration of cathode rays into a gas produces in it negative ions identical with those of Roentgen rays, at least from the point of view of their mobility or of their power of condensing supersaturated water vapor. It seems, nevertheless, important to make sure, by measuring the mobility of ions produced by different causes in the interior of gases, whether the differences which appear to exist are real and are caused by the difference in the molecules which adhere to them, or are due to the electrified centres which serve as the nuclei for them.

(44) *Mobility and Recombination.* It is equally important to be able, by measurement of mobility, to follow the modification which a change of temperature produces in the size of the agglomeration and to connect the ions observed at ordinary temperatures with the incomparably more mobile ions which we observe in flames, and which appear to be made up of single electrical centres, cathode corpuscles and perhaps *a* particles.

The rate of recombination of ions is as yet not well known in respect to the variations with pressure and temperature, although it certainly plays an essential part in the phenomena of disruptive discharge through gases at low pressures; it would be desirable if this point were better fixed.

(45) *Ionization by Impact.* Every actual theory of the disruptive discharge rests on the conception that the impact of an electrified particle in sufficiently rapid motion against a molecule can cause corpuscular dissociation.

This idea was a natural consequence of the known fact that cathode and Becquerel rays, made up of similar particles, make a gas through which they pass a conductor. If the corpuscular dissociation produces in the gas, separated from the molecule, a cathode corpuscle and a positive residue, these fragments can, if a sufficiently intense electric field exists in the gas, acquire a velocity great enough to act as β or *a* rays and cause from point to point a rapid increase in conductivity.

Townsend has shown how this consequence is capable of exact experimental verification, and he has found that between certain limits of velocity, each impact between the cathode corpuscle and a molecule results in a corpuscular dissociation of the same kind. The velocity acquired ought not, however, to exceed a certain limit beyond which the negative corpuscle or β particle passes through the atomic edifice without producing a sensible disturbance in it.

In order that a disruptive discharge may exist without an external cause to maintain the production of the first electrified centres, it is necessary that the positive centres should be able, like the negative, although with more difficulty, to produce corpuscular dissociation at the moment of their impact with the molecules, as this latter causes the conductivity produced in gases by the *a* rays.

Townsend has been able, in support of this hypothesis, to determine the exact moment when the disruptive phenomenon is produced, and to analyze the mechanism of it.

In addition to this fundamental conception of ionization by impact, the theory of the disruptive discharge has yet much progress to make.

The extremely varied aspects which this discharge takes, the production of striations, an explanation of which was first given by J. J. Thomson, the influence of a magnetic field on the conditions of the discharge, the phenomena that are produced when the electrodes are only of the order of a micron apart, where the molecules do not appear to take part in the production of the spark, are many of the essential points which to-day attract attention.

(46) *The Electric Arc.* By the side of the ordinary disruptive discharge, by brush or spark, the electric arc, with an entirely different aspect, brings in the new phenomenon of the emission of cathode corpuscles by the surface of incandescent bodies. This incandescence of the electrode, of the cathode especially, is, in fact, characteristic of the arc discharge; the cathode is raised to a sufficiently high temperature by the impact of the positive ions which flow toward it, so that the corpuscles present in the electrode, and which give it its conductivity, experience a true evaporation and carry the greater part of the current. In fact, a filament of incandescent carbon is able to emit, at a much lower temperature than that of the voltaic arc, cathode corpuscles representing a current density of two amperes per square centimeter.

(47) *Evaporation of the Cathode.* This phenomenon, known under the name of the Edison effect, is very general and has been connected in a quantitative manner by Richardson on the fundamental hypothesis of the kinetic theory with the presence of freely moving cathode particles in the interior of conductors.

At ordinary temperatures this emission of corpuscles is diminished to such an extent that electrostatics is possible and a metal can keep a permanent charge. Every corpuscle present in the metal is immersed in a medium of high specific inductive capacity, and a finite amount of work is necessary to make them pass from this medium to a region where the specific inductive capacity is equal to unity. Only the corpuscles having a sufficient velocity would be able to supply this work on leaving the conductor, and their number, absolutely negligible at ordinary temperatures, increases with extreme rapidity with the rise in temperature. Richardson has shown that the variation obtained by experiment agrees very well with that predicted by theory.

(48) *Metals.* The spontaneous dissociation of atoms which the kinetic theory implies, the separation of electrified centres free to move in the interior of the metal, is a consequence of the high specific inductive capacity of the medium, of the ease of electrostatic polarization of metals, owing to the ease with which the metallic atoms lose corpuscles in order to remain positively charged. The potential energy of an electrified particle in such a medium is much smaller than anywhere else, and conformably with the laws of the distribution of

energy given by the kinetic theory, the free particles ought to be more numerous in it.

(49) *Chemical Phenomena.* It is by an action of the same kind that water, of great specific inductive capacity (smaller, however, than that of metals) causes the electrolytic dissociation of salts that are dissolved in it; it would be of great interest to determine the relation between this electrolytic dissociation, especially of liquid conductors, and the corpuscular dissociation common probably to gases and metals.

In electrolytic dissociation, the cathode corpuscles lost by the metallic atoms, instead of remaining free as in corpuscular dissociation, remain united to an atom or to a radical to form the negative ion in electrolytes. This question touches the relations between our actual ideas and chemistry, relations still very obscure, and which it would be very important to clear up. The electric dissociation produced in gases by Roentgen rays does not appear connected with any chemical modification; however, in air all intense ionization is accompanied by the formation of ozone. Here is a domain almost entirely unexplored.

X. *Magnetic Properties*

(50) *Ampère and Weber.* However, the complex phenomena of magnetism and diamagnetism have seemed so far to lead us to expect more difficulties, although the electrons gravitating in the atom in closed orbits furnish at first sight a simple representation of the molecular currents of Ampère, capable of turning under the action of an external magnetic field in order to give birth to induced magnetism, or of reacting by induction, according to the idea of Weber, against the external field so as to make the substance diamagnetic.

Those who have tried to follow out this idea have found it so far sterile; independently, different physicists have come to the conclusion that the hypothesis of electrons in undiminished motion cannot furnish a representation of the permanent phenomena of magnetism or diamagnetism.

I am enough of a parvenu to attempt to show, contrary to the preceding opinion, that it is possible to give, by means of the electrons, an exact signification to the ideas of Ampère and Weber, to find for para- and diamagnetism completely distinct interpretations, conforming to the laws experimentally established by Curie: weak magnetism, an attenuated form of ferromagnetism, varies inversely as the absolute temperature; on the other hand diamagnetism is shown to be, in all observed cases with the exception of bismuth, rigorously independent of the temperature. The theory which I propose takes entire account of these facts and clears up at the same time the complex question of magnetic energy.

I shall give here only the principal results of this work which will, be published in full elsewhere.

(51) *Molecular Currents.* An electrified particle of charge e moving with a velocity v is equivalent to a current of moment ev. One easily deduces from this that a molecular current made up of an electron which describes in the periodic time t an orbit inclosed by the surface S is equivalent from the point of view of the magnetic field produced to a magnet of magnetic moment $M = \frac{eS}{t}$ normal to the plane of the orbit.

There would be a corresponding current for each of the electrons present in a molecule, and the magnetic moment resulting from these would be zero or different from zero, according to the degree of symmetry of the molecular structure.

(52) *Diamagnetism.* If on a group of such molecules we superimpose an external magnetic field, all the molecular currents experience a modification independent of the manner in which the superposition is obtained, whether by the establishment of the field or by motion of the molecule in a preëxisting field. The direction of this modification, due to the induction experienced by the molecular currents, corresponds always to diamagnetism, the increase of the magnetic moment being $\Delta M = -\frac{He^2}{4\pi m} S$ in the case of a circular orbit. H is the component of the magnetic field normal to the plane of the orbit and m the mass of the electron which describes the orbit.

(53) *The Magnetic Energy.* When the molecule is supposed immovable, the work necessary for the modification of the molecular currents is furnished by the electric field produced, according to the equations of Hertz, during the establishment of the magnetic field.

In the opposite case, where the modification is due to the motion of the molecules, the work is furnished to the molecular currents by the kinetic energy of the molecule or by the action of neighboring molecules. The diamagnetic modification produced at the moment of the establishment of the field continues in spite of the molecular agitation.

This modification is manifested in three distinct ways:

1. If the resulting motion of the molecules is zero, the substance is diamagnetic in the ordinary sense of the word, and the order of magnitude of the experimental diamagnetic constants is in good agreement with the hypothesis of molecular currents circulating in intra-molecular paths.

This conception leads to the law of independence established by Curie between the diamagnetic constants and the temperature or the physical state.

2. If the resulting motion of the molecules is not zero, the initial diamagnetic modification is followed by an orientation of the molecules under the action of the external field, which cause a paramagnetism to appear that masks the underlying diamagnetism, the new phenomenon being considerable compared to the first, when the symmetry permits it to appear.

In slightly paramagnetic bodies, such as gases, the heat agitation is opposed to the complete orientation of the molecular magnets, to saturation, and one finds, in seeking what permanent condition is established, the law of Curie, that the variation of paramagnetic constants is in inverse ratio to the absolute temperature.

3. Finally, the change of period of revolution in consequence of the diamagnetic modification corresponds to the Zeeman effect, as general as diamagnetism itself; iron, certain rays of which show the Zeeman effect, is diamagnetic before the orientation of the molecular magnets under the action of the external field makes it appear paramagnetic.

The orbits considered, which represent the molecular currents of Ampère, are also the circuits of zero resistance of the diamagnetism of Weber, with this remarkable peculiarity that the flux which passes through them is not constant, as Weber supposed, if the inertia of the electrons is entirely of electromagnetic origin.

I have shown, on the other hand, that the orbits of the electrons supposed circular, and described under the action of central forces, experience no deformation during the diamagnetic modification, this latter consisting only in a change of velocity of the electrons in their orbits. We can thus form an exact and simple conception of the facts of magnetism and diamagnetism by considering the molecular currents as non-deformable but movable currents, of zero resistance and of enormous self-induction, to which all the ordinary laws of induction are applicable.

XI. *Conclusion*

The rapid perspective which I have just sketched is full of promises, and I believe that rarely in the history of physics has one had the opportunity of looking either so far into the past or so far into the future. The relative importance of parts of this immense and scarcely explored domain appears different to-day from what it did in the preceding century: from the new point of view the various plans arrange themselves in a new order. The electrical idea, the last discovered, appears to-day to dominate the whole, as the place of choice where the explorer feels that he can found a city before advancing into new territories.

The mechanical facts, the most evident of all those of which matter is possessed, from the first attracted the attention of our ancestors, and led them to conceive of the notions of mass and force which appeared a long while the most fundamental, those from which all the others ought to raminate. As the means of investigation have increased, as the more hidden facts have been discovered, we have thought for a long while to be able to reduce them to the old laws, to be able in fact to find an explanation of mechanical origin.

The actual tendency, of making the electromagnetic ideas to occupy the preponderating place, is justified, as I have sought to show, by the solidity of the double base on which rests the idea of the electron; on the one hand by the exact knowledge of the electro-magnetic ether which we owe to Faraday, Maxwell, and Hertz, and on the other hand by the experimental evidence brought forward by the recent investigations into the granular structure of electricity. Moreover, this assurance which we express when considering the past is increased, if it is possible, when we consider the future.

Already all views, not only of the ether, but of matter, source and receiver of luminous waves, obtain an immediate interpretation which mechanics is powerless to give, and this mechanics itself appears to-day as a first approximation, largely sufficing in all cases of motion of matter taken in mass, but for which a more complete expression must be sought in the dynamics of the electron.

Although still very recent, the conceptions of which I have sought to give a collected idea are about to penetrate to the very heart of the entire physics, and to act as a fertile germ in order to crystallize around it, in a new order, facts very far removed from one another.

Falling in ground well prepared to receive it, in the ether of Fara-day, Maxwell, and Hertz, the idea of the electron, an electrified movable centre which experiment to-day allows us to lay hold of individually, constitutes the tie between the ether and matter formed of a group of electrons.

This idea has taken an immense development in the last few years, which causes it to break the framework of the old physics to pieces, and to overturn the established order of ideas and laws in order to branch out again in an organization which one foresees to be simple, harmonious, and fruitful.

Ernest Rutherford
1871–1937

PRESENT PROBLEMS OF RADIOACTIVITY

BY ERNEST RUTHERFORD

[Ernest Rutherford, Macdonald Professor of Physics, McGill University, Montreal. b. August 30, 1871, Nelson, New Zealand. M.A. and D.Sc. University of New Zealand; B.A. Cambridge, England; F.R.S. 1903; F.R.S.C. 1899; postgraduate, Cambridge, England, 1895–98; Professor of Physics, McGill University, Montreal, 1898. Member of American Physical Society, and others; awarded Rumford Medal of the Royal Society, 1904. **Author of** books and articles.]

I

SINCE the initial discovery by Becquerel of the spontaneous emission of new types of radiation from uranium, our knowledge of the phenomena exhibited by uranium and the other radioactive bodies has grown with great and ever increasing rapidity, and a very large mass of experimental facts has now been accumulated. It would be impossible within the limits of this article even to review briefly the more important experimental facts connected with the subject, and, in addition, such a review is rendered unnecessary by the recent publication of several treatises [1] in which the main facts of radioactivity have been dealt with in a fairly complete manner.

In the present article, an attempt will be made to discuss the more important problems that have arisen during the development of the subject and to indicate what, in the opinion of the writer, are the subjects which will call for further investigation in the immediate future.

II. *Nature of the radiations*

The characteristic radiations from the radioactive bodies are very complex, and a large amount of investigation has been necessary to isolate the different kinds of rays and to determine their specific character. The rays from the three most studied radio-elements, uranium, thorium, and radium, can be separated into three distinct types, known as the α, β, and γ rays.

The nature of the α and β rays has been deduced from observations of the deflection of the path of the rays by a magnetic and electric field. According to the electromagnetic theory, a radiation which is deflectable by a magnetic or electric field must consist of a flight of charged particles. If the amount of deflection of the rays from their path is measured when both a magnetic and an electric field of known

---

[1] Mme. Curie, *Thèses présentées à la Faculté des Sciences.* Paris, 1903.
H. Becquerel, *Recherches sur une propriété nouvelle de la Matière.* Typographie de Firmin, Didot et Cie. Paris, 1903.
E. Rutherford, *Radioactivity.* Cambridge, University Press, 1904.
F. Soddy, *Radioactivity.* Electrician Co., London, 1904.

strength is applied, the value V of the velocity of the particles and the ratio $\frac{e}{m}$ of the charge carried by the particle to its apparent mass m can be determined. From the direction of the deviation, the sign of the electric charge carried by the particle can be deduced.

Examined in this way, the β rays have been shown to consist of negatively charged particles projected with a velocity approaching that of light. The experiments of Becquerel and Kaufmann have shown that the β rays are identical with the cathode rays produced in a vacuum tube. This relationship has been established by showing that the value of $\frac{e}{m}$ is the same for the two kinds of rays. In both cases the value of $\frac{e}{m}$ has been found to be about 10^7 electromagnetic units, while the corresponding value of $\frac{e}{m}$ for hydrogen atoms set free in the electrolysis of water is 10^4. If the charge on the β particles — or electrons as they may be termed — is the same as that carried by the hydrogen atom, this result shows that the apparent mass of the electrons at slow speeds is about $\frac{1}{1000}$ of that of the hydrogen atom. The β particles from the radio-elements are expelled with a much greater speed than the cathode ray particles in a vacuum tube. The velocity of the β particles from radium is not the same for all particles, but varies between about 10^{10} and 3×10^{10} cms. per second. The swifter particles move with a velocity of at least 95 per cent of that of light. The emission by radium of electrons with high but different velocities has been utilized by Kaufmann to determine the variation of $\frac{e}{m}$ with speed. He found that the value of $\frac{e}{m}$ decreased with increase of velocity, showing that the apparent mass increased with the speed. By comparison of the experimental results with the mathematical theory of a moving charge, he deduced that the mass of the electrons was in all probability electromagnetic in origin, *i. e.*, the apparent mass could be explained purely in terms of electricity in motion without the necessity of a material nucleus on which the charge was distributed. J. J. Thomson, Heaviside, and others, have shown that a moving charged sphere increases in apparent mass with the speed, and that, for speeds small compared with the velocity of light, the increase of mass $m = \frac{2}{3} \frac{e^2}{a}$ where e is the charge carried by the body and a the radius of the conducting sphere over which the electricity is distributed. Kaufmann deduced that the value of $\frac{e}{m} = 1.86 \times 10^7$ for electrons of slow velocity. If the mass of the electrons is electrical in origin, it is seen that $a = 10^{-13}$ cms., since the value of $e = 3.4 \times 10^{-10}$ electrostatic units. The results of various methods of determination agree in fixing the diameter of an atom as about 10^{-8} cms. The apparent diameter of an electron is thus minute compared with that of the atom itself.

The highest velocity of the radium electrons measured by Kaufmann was 95 per cent of the velocity of light. The power of electrons of penetrating solid matter increases rapidly with the velocity, and some of those expelled from radium are able to penetrate through more than 3 mms. of lead. It is probable that a few of the electrons from radium move with a velocity still greater than the highest value observed by Kaufmann, and it is important to determine the value of $\frac{e}{m}$ and the velocity of such electrons. According to the mathematical theory, the mass of the electron increases rapidly as the speed of light is approached, and should be infinitely great when the velocity of light is reached. This leads to the conclusion that no charged body can be made to move with a velocity greater than that of light. This result is of great importance, and requires further experimental verification. A close study of the high speed electrons from radium may throw further light on this question.

Only a brief and imperfect statement of our knowledge of electrons has been given in this paper. A more complete and detailed account of both the theory and experiment will be given by my colleague, Dr. Langevin.

III. *The a rays*

The β rays are readily deflected by a magnetic field, but a very intense magnetic field is required to deflect appreciably the a rays. The writer showed by the electric method that the a rays of radium were deflected both by a magnetic and electric field, and deduced the velocity of projection of the particles and the ratio $\frac{e}{m}$ of the charge to the mass. The direction of deflection of the a rays is opposite in sense to the β rays. Since the β rays carry a negative charge, the a particles thus behave as if they carried a positive charge. The magnetic deflection of these rays was confirmed by Becquerel and Des Coudres, using the photographic method, while the latter, in addition, showed their deflection in an electric field and deduced the value of the velocity and $\frac{e}{m}$. The values obtained by Rutherford and Des Coudres were in very good agreement, considering the difficulty of obtaining a measurable deviation.

Observer	Value of Velocity	Value of $\frac{e}{m}$
Rutherford	2.5×10^9 cms. per sec.	6×10^3 electromagnetic units
Des Coudres	1.6×10^9 cms. per sec.	6×10^3 electromagnetic units

Since the value of $\frac{e}{m}$ for the hydrogen atom is 10^4, on the assumption that the a particle carries the same charge as the hydrogen atom, this result shows that the apparent mass of the a particle is about twice that of the hydrogen atom. If the a particle consists of any known kind of matter, this result indicates that it is either the atom of hydrogen or of helium. The a particles thus consist of heavy bodies projected with great velocity, whose mass is of the same order of magnitude as the helium atom and at least 2000 times as great as the apparent mass of the β particle.

If the a particles carry a positive charge, it is to be expected that the particles, falling on a body of sufficient thickness to absorb them, should under suitable conditions give it a positive charge, while the substance from which they are projected should acquire a negative charge. The corresponding effect has been observed for the β rays. The β particles from radium communicate a negative charge to the body on which they fall, while the radium from which they are emitted acquires a positive charge. This effect has been very strikingly shown by a simple experiment of Strutt. The radium compound, sealed in a small glass tube, the outer surface of which is made conducting, is insulated by a quartz rod. A simple gold-leaf electroscope is attached to the bottom of the glass tube, in order to indicate the presence of a charge. The whole apparatus is inclosed in a glass vessel, which is exhausted to a high vacuum, in order to reduce the loss of charge in consequence of the ionization of the gas by the rays. Using a few milligrams of radium bromide, the gold leaf diverges to its full extent in a few minutes and shows a positive charge. The explanation is simple. A large proportion of the negatively charged particles are projected through the glass tube containing the radium, and a positive charge is left behind. By allowing the gold leaf, when extended, to touch a conductor connected to earth, the gradual divergence of the leaves and their collapse becomes automatic, and will continue, if not indefinitely, at any rate for as long a time as the radium lasts.

When the radium is exposed under similar conditions, but unscreened in order to allow the a particles to escape, no such charging action is observed. This is not due to the equality between the number of positively and negatively charged particles expelled from the radium, for no effect is observed when the radium is temporarily freed from its power of emitting β rays by driving off the emanation by heat. The writer recently attempted to detect the charge carried by the a rays from radium by allowing them to fall on an insulated plate in a vacuum, but no appreciable charging was observed. The β rays were temporarily got rid of by heating the radium in order to drive off its emanation. There was found to be a strong surface ionization set up at the surface from which the rays emerged and the surface on which they impinged. The presence of this ionization causes the upper plate to lose rapidly a charge communicated to it. Although this action would mask to some extent the effect to be looked for, a measurable effect should have been obtained under the experimental conditions, if the a rays were expelled with a positive charge; but not the slightest evidence of a charge was observed. I understand that similar negative results have been obtained by other observers.

This apparent absence of charge carried by the a rays is very remarkable and difficult to account for. There is no doubt that the a particles *behave* as if they carried a positive charge, for several observers have shown that the a rays are deflected by a magnetic field. It is interesting, in this connection, that Wien was unable to detect that the "canal rays" carried a charge. These rays, discovered by Goldstein, are analogous in many respects to the a rays. They are slightly deflected by a magnetic and electric field, and behave like positively charged bodies atomic in size. The value of $\frac{e}{m}$ is not a constant, but depends upon the nature of the gas in the tube through which the discharge is passed. The apparent absence of charge on the a particles may possibly be explained on the supposition that a negatively charged particle (an electron) is always projected at the same time as the positively charged particle. Such electrons if they are present should be readily bent back to the surface from which they came by the action of a strong magnetic field. It will be of interest to examine whether the charge carried by the a rays can be detected under such conditions. Another hypothesis, which has some points in its favor, is that the a particles are uncharged at the moment of their expulsion, but, in consequence of their collision with the molecules of matter, lose a negative electron, and consequently acquire a positive charge. This point is at present under examination. The question is in a very unsatisfactory state, and requires further investigation.

It is remarkable that positive electricity is always associated with matter atomic in size, for no evidence has been obtained of the existence of a positive electron corresponding to the negative electron. This difference between positive and negative electricity is apparently fundamental, and no explanation of it has, as yet, been forthcoming.

The evidence that the a particles are atomic in size mainly rests on the deflection of the path of the rays in a strong magnetic and electric field. It has, however, been suggested by H. A. Wilson that the a particle may in reality be a "positive" electron, whose magnitude is minute compared with that of the negative. The electric mass of an electron for slow speeds is equal to $\frac{2}{3}\frac{e^2}{a}$. Since there is every reason to believe that the charge carried by the a particle and the electron is the same, in order that the mass of the positive electron should be about 2000 times that of the negative, it would be necessary to suppose that the radius of the sphere over which the charge is distributed is only $\frac{1}{2000}$ of that of the electron, *i. e.*, about 10^{-16} cms. The magnetic and electric deflection would be equally well explained on this view. This hypothesis, while interesting, is too far-reaching in its consequences to accept before some definite experimental evidence is forthcoming to support it. The evidence at present obtained strongly supports the view that the a particles are in reality projected matter, atomic in size. The probability that the a particle is an atom of helium is discussed later, in section VIII.

Becquerel showed that the a rays of polonium were deflected by a magnetic field to about the same extent as the a rays of radium. On account of the feeble activity of thorium and uranium, compared with radium and polonium, it has not been found possible to examine whether the a rays emitted by them are deflectable. There is little doubt, however, that the a particles of all the radio-elements are projected matter of the same kind (probably helium atoms). The a rays from the different radioactive products differ in their power of penetration of matter in the proportion of about three to one, being greatest for the a rays from the imparted or "induced" activity of radium and thorium, and least for uranium. This difference is probably mainly due to a variation of the velocity of projection of the a particles in the various cases. The interpretation of results is rendered difficult by our ignorance of the mechanism of absorption of the a rays by matter. Further experiment on this point is very much required.

It is of importance to settle whether the a particles of radium and polonium have the same ratio of $\frac{e}{m}$. Becquerel states that the amount of curvature of the a rays from polonium in a field of constant strength was the same as for the a rays from radium. This would show that the product of the mass and velocity is the same for the a particles from the two substances. The a rays of polonium, however, certainly have less penetrating power than those of radium, and presumably a smaller velocity of projection. This result would indicate that $\frac{e}{m}$ is different for the a particles of polonium and radium. It is of importance to determine accurately the ratio of $\frac{e}{m}$ and the velocity for the rays from these two substances in order to settle this important point.

IV. *The γ Rays*

In addition to the a and β rays, uranium, thorium, and radium all emit very penetrating rays known as γ rays. These rays are about 100 times as penetrating as the β rays, and their presence can be detected after passing through several centimeters of lead. Villard, who originally discovered these rays in radium, stated that they were not deflected in a magnetic field, and this result has been confirmed by other observers. Quite recently, Paschen has described some experiments which led him to believe that the γ rays are corpuscular in character, consisting of negatively charged particles (electrons) projected with a velocity very nearly equal to that of light. This conclusion is based on the following evidence. Some pure radium bromide was completely inclosed in a lead envelope 1 cm. thick, — a thickness sufficient to absorb completely the ordinary β rays emitted by radium, but which allows about half of the γ rays to escape. The lead envelope was insulated in an exhausted vessel, and was found to gain a positive charge. In another experiment, the rays escaping from the lead envelope fell on an insulated metal ring, surrounding the lead envelope. When the air was exhausted, this outer ring was found to gain a negative charge. These experiments, at first sight, indicate that the γ rays carry with them a negative charge like the β rays. In order to account for the absence of deflection of the path of the γ rays in very strong magnetic or electric fields, it is necessary to suppose that the particles have a very large apparent mass. Paschen supposes that the γ particles negative are electrons like the β particles, but are projected with a velocity so nearly equal to that of light that their apparent mass is very great.

Some experiments recently made by Mr. Eve, of McGill University, are of great interest in this connection. He found by the electric method that the γ rays set up secondary rays, in all directions, at the surface of which they emerge and also on the surface of which they impinge. These rays are of much less penetrating power than the primary rays, and are readily deflected by a magnetic field. The direction of deflection indicated that these secondary rays consisted, for the most part, of negatively charged particles (electrons) projected with sufficient velocity to penetrate through about 1 mm. of lead. In the light of these results, the experiments of Paschen receive a simple explanation without the necessity of assuming that the γ rays of radium themselves carry a negative charge. The lead envelope in his experiment acquired a positive charge in consequence of the emission of a secondary radiation consisting of negatively charged particles, projected with great velocity from the surface of the lead. The electric charge acquired by the metal ring was due to the absorption of these secondary rays by it, and the diminution of this charge in a magnetic field was due to the ease with which these secondary rays are deflected. It is thus to be expected that the envelope surrounding the radium, whether made of lead or other metal, would always acquire a positive charge, provided the metal is not of sufficient thickness to absorb all the γ rays in their passage through it.

No conclusive evidence has yet been brought forward to show that the γ rays can be deflected either in a magnetic or electric field. In this, as in other respects, the rays are very analogous to the Roentgen X rays.

According to the theory of Stokes, J. J. Thomson, and Weichert, of rays are transverse pulses set up in the ether by the sudden arrest X the motion of the cathode particles on striking an obstacle. The more sudden the stoppage, the shorter is the pulse, and the rays, in consequence, have greater power of penetrating matter. In some recent experiments Barkla found that the secondary rays set up by the X rays, on striking an obstacle, vary in intensity with the orientation of the X-ray tube, showing that the X rays exhibit the property of one-sidedness or polarization. This is the only evidence so far obtained in direct support of the wave-nature of the X rays.

If X rays are not set up when the cathode particles are stopped, conversely, it is to be expected that X rays should be set up when they are suddenly set in motion. Now this effect is not observable in an X-ray tube, since the cathode particles acquire most of their velocity, not at the cathode itself, but in passing through the electric field between the cathode and anti-cathode. It is, however, to be expected theoretically that a type of X rays should be set up at the sudden expulsion of the β particles from the radio-atoms. The rays, too, should be of a very penetrating kind, since not only is the charged particle projected with a speed approaching that of light, but the change of motion must occur in a distance comparable with the diameter of an atom.

On this view, the γ rays are a very penetrating type of X rays, having their origin at the moment of the expulsion of the β particle from the atom. If the β particle is the parent of the γ rays, the intensity of the β and γ rays should, under all conditions, be proportional to one another. I have found this to be the case, for the γ rays always accompany the β rays and, in whatever way the β-ray activity varies, the activity measured by the γ rays always varies in the same proportion. Active matter which does not emit β rays does not give rise to γ rays. For example, the radio-tellurium of Marckwald, which does not emit β rays, does not give off γ rays.

Certain differences are observed, however, in the ionizing action of γ and X rays. For example, gases and vapors like chlorine, sulphuretted hydrogen, methyl-iodine, and chloroform, when exposed to ordinary X rays, show a much greater ionization, compared with air, than is to be expected, according to the density law. On the other hand, the relative ionization of these substances by γ rays follows the density law very closely. It seemed likely that this apparent difference was due mainly to the greater penetrating power of the γ rays. This was confirmed by some recent experiments of Eve, who found that the relative conductivity of gases exposed to very penetrating X rays from a hard tube approximated in most cases closely to that observed for the γ rays. The vapor of methyl-iodine was an exception, but the difference in this case would probably disappear if

X rays could be generated of the same penetrating power as that of the γ rays.

The results so far obtained thus generally support the view that the γ rays are a type of penetrating X rays. This view is in agreement, too, with theory, for it is to be expected that very penetrating β rays should always appear with the β rays.

No evidence of the emission of a type of X rays is observed from active bodies which emit only a rays. If the a particles are initially projected with a positive charge, such rays are to be expected. Their absence supplies another piece of evidence in support of the view that the a particle is projected without a charge, but acquires a positive charge in its passage through matter.

V. *Emission of Energy by the Radioactive Bodies*

It was early recognized that a very active substance like radium emitted energy at a rapid rate, but the amount of this energy was very strikingly shown by the direct measurements of its heating effect made by Curie and Laborde. They found that one gram of radium in radioactive equilibrium emitted about 100 gram calories of heat per hour. A gram of radium would thus emit 896,000 gram calories per year, or over 200 times as much heat as is liberated by the explosion of hydrogen and oxygen to form one gram of water. They showed that the rate of heat emission was the same in solution as in the solid state, and remained constant when once the radium had reached a stage of radioactive equilibrium. Curie and Dewar showed that the rate of evolution of heat from radium was unaltered by plunging the radium into liquid air, or liquid hydrogen.

It seemed probable that the evolution of heat by radium was directly connected with its radioactivity, and the experiments of Rutherford and Barnes proved this to be the case. The heating effect of a quantity of radium bromide was first determined. The emanation was then completely driven off by heating the radium, and condensed in a small glass tube by means of liquid air. After removal of the emanation, the heat evolution of the radium in the course of about three hours fell to a minimum corresponding to one quarter of its original value, and then slowly increased again, reaching its original value after an interval of about one month. The heat emission from the emanation tube at first increased with the time, rising to a maximum value about three hours after its introduction. It then slowly decreased according to an exponential law with the time, falling to half value in about four days. If Q_{max} is the maximum heating effect of the emanation tube, the heat emission Q at any time t, after the maximum is reached, is given by

$$Q = Q_{max} e^{-\lambda t}$$

where λ is the radioactive constant of the emanation.

The curve expressing the recovery with time of the heating effect of radium from its minimum is complementary to the curve expressing the diminution of the heating effect of the emanation tube with time. The curves of decay and recovery agree within the limit of experimental error with the corresponding curves of decay and recovery of the activity of radium when measured by the a rays. Since the minimum, or non-separable activity of radium, measured by the a rays, after the emanation has been removed, is only one quarter of the maximum activity, these results indicate that the heating effect of radium is proportional to its activity measured by the a rays. It is not proportional to the activity measured by the β or γ rays, since the β or γ ray activity of radium almost completely disappears some hours after removal of the emanation.

These results have been confirmed by further observations of the distribution of the heat emission between the emanation and the successive products which arise from it. If the emanation is left for several hours in a closed tube, its activity measured by the electric method increases to about twice its initial value. This is due to the "excited activity," or in other words to the radiations from the active matter deposited on the walls of the tube by the emanation. The activity of this deposit has been very carefully analyzed, and the results show that the matter deposited by the emanation breaks up in three successive and well-marked stages. For convenience, these successive products of the emanation will be termed radium A, radium B, and radium C. The time T taken for each of these products to be half transformed, and the radiations from each product, are shown in the following table:

Product	T	Radiations
Radium		a rays
$\downarrow$		
Emanation	4 days	a rays
$\downarrow$		
Radium A	3 mins.	a rays
$\downarrow$		
Radium B	21 mins.	no rays
$\downarrow$		
Radium C	28 mins.	a, β, and γ rays

When the emanation has been left in a closed vessel for several hours, the emanation and its successive products reach a stage of approximate radioactive equilibrium, and the heating effect is then a maximum. If the emanation is suddenly removed from the tube by a current of air, the heating effect is then due to radium A, B, and C together. On account, however, of the rapidity of the change of radium A (half value in three minutes), it is experimentally very difficult to distinguish between the heating effect of the emanation and that of radium A. The curve of variation with time of the heating effect of the emanation tube after removal of the emanation is very nearly the same as the corresponding curve for the activity measured by the a rays. These results show that each of the products of radium supplies an amount of heat roughly proportional to its activity measured by the a rays. Each product loses its heating effect at the same rate as it loses its activity, showing that the heating effect is directly connected with the radioactive changes. The results indicated that the product, radium B, which does not emit rays does not supply an amount of heat comparable with the other products. This point is important, and requires more direct verification.

Since the heat emission is in all cases nearly proportional to the number of a particles expelled, the question arises whether the bombardment of these particles is sufficient to account for the heating effects observed. The kinetic energy of the a particle $\frac{1}{2}mv^2$ can be at once determined since $\frac{e}{m}$ and V are known. The following table shows the kinetic energy of the a particle deduced from the measurements of Rutherford and Des Coudres. The third column shows the number of a particles expelled from 1 gram of radium per second on the assumption that the heating effect of radium (100 gram-calories per gram per hour) is entirely due to the energy given out by the expelled a particles.

Observer	Kinetic energy	Number of a particles expelled per second from 1 gram of radium.
Rutherford	5.9×10^{-6} ergs.	2×10^{-11}
Des Coudres	2.5×10^{-6} ergs.	5×10^{-11}

This hypothesis that the heating effect of radium is due to bombardment of the a particle can be indirectly put to the test in the following way. It seems probable that each atom of radium in breaking up emits one a particle. On the disintegration theory, the residue of the atom, after the a particle is expelled, is the atom of the emanation, so that each atom of radium gives rise to one atom of the emanation. Let q be the number of atoms in each gram of radium breaking up per second. When a state of radioactive equilibrium is reached, the number N of emanation particles present is given by $N = \frac{q}{\lambda}$, where λ is the constant of change of the emanation. Now Ramsay and Soddy deduced from experiment that the volume of the emanation released from 1 gram of radium was about one cubic millimeter at atmospheric pressure and temperature. It has been experimentally deduced that there are 3.6×10^{19} molecules in one cubic centimeter of gas at ordinary pressure and temperature. The emanation obeys Boyle's law and behaves, in all respects, like a heavy gas, and we may in consequence deduce that $N = 3.6 \times 10^{16}$. Now $\lambda = 2.0 \times 10^{-6}$. Thus $q = 7.2 \times 10^{10}$. Now the particles expelled from radium in a state of radioactive equilibrium are about equally divided between four substances, viz., the radium itself, the emanation, radium A and C. We may thus conclude that the number of a particles expelled per second from 1 gram of radium in radioactive equilibrium is 2.9×10^{11}. The value deduced by this method is intermediate between the values previously obtained (see previous table) on the assumption that the heating effect is entirely due to the a particles.

I think we may conclude from the agreement of these two methods of calculation that the greater portion of the heating effect of radium is a direct result of the bombardment of the expelled a particles, and that, in all probability, about 5×10^{10} atoms of radium break up per second.

The energy carried off in the form of β and γ rays is small compared with that emitted in the form of a rays. By calculation it can be shown that the average kinetic energy of the β particle is small in comparison with that of the a particle. This result is confirmed by comparative measurements of the total ionization produced by the a and β rays, when the energy of the rays is all used up in ionizing the gas, for the total ionization produced by the β rays is small compared with that due to the a rays. The total ionization produced by the γ rays is about the same as that produced by the β rays, showing that, in all probability, the energy emitted in the form of these two types of radiation is about the same. From the point of view of the energy radiated, and of the changes which occur in the radioactive bodies, the a rays thus play a far more important rôle in radioactivity than the β or γ rays. Most of the products which arise from radium and thorium emit only a rays, while the β and γ rays appear only in the last of the series of rapid changes which take place in these bodies.

Since most of the heating effect of radium is due to the a rays, it is to be expected that all radioactive substances, which emit a rays, should also emit heat at a rate proportional to their a ray activity. On this view, both uranium and thorium should emit heat at about one millionth the rate of radium. It is of importance to determine directly the heating effect for these substances, and also for actinium radio-tellurium.

According to the disintegration theory, the a particle is expelled as a result of the disintegration of the atom of radioactive matter. While it is to be expected that a greater portion of the energy emitted should be carried off in the form of kinetic energy by the expelled particles, it is also to be expected that some energy would be radiated in consequence of the rearrangement of the components of the system after the violent ejection of one of its parts. No direct measurements have yet been made of the heating effect of the a particles, independently of the substance in which they are produced. Experiments of this character would be difficult, but would throw light on the important question of the division of the energy radiated between the expelled a ray particle and the system from which it arises.

The enormous evolution of energy by the radioactive substances is very well illustrated by the case of the radium emanation. The emanation released from 1 gram of radium in radioactive equilibrium emits during its changes an amount of energy corresponding to about 10,000 gram-calories. Now Ramsay and Soddy have shown that the volume of this emanation is about 1 cubic millimeter at standard pressure and temperature. One cubic millimeter of the emanation and its product thus emits about 10^7 gram-calories. Since 1 centimeter of hydrogen, in uniting with the proportion of oxygen required to form water, emits 3.1 gram-calories, it is seen that the emanation emits about 3 million times as much energy as an equal volume of hydrogen.

It can readily be calculated, on the assumption that the atom of the emanation has a mass 100 times that of hydrogen, that 1 pound of the emanation some time after removal could emit energy at the rate of about 8000 horse-power. This would fall off in a geometrical progression with the time, but, on an average, the amount of energy emitted during its life corresponds to 50,000 horse-power days. Since the radium is being continuously transformed into emanation, and three quarters of the total heat emission is due to the emanation and its products, a simple calculation shows that 1 gram of radium must emit during its life about 10^9 gram-calories. As we have seen, the heat emission of radium is about equally divided between the radium itself and the three other a ray products which come from it.

The heat emitted from each of the other radioactive substances while their activity lasts, should be of the same order of magnitude, but in the case of uranium and thorium the present rate of heat emission would probably continue, on an average, for a period of about 1000 million years.

VI. *Source of the Energy emitted by the Radioactive Bodies*

There has been considerable difference of opinion in regard to the fundamental question of the origin of the energy spontaneously emitted from the radioactive bodies. Some have considered that the atoms of the radio-elements act as transformers of borrowed energy. The atoms are supposed to be able, in some way, to abstract energy from the surrounding medium and to emit it again in the form of the characteristic radiations observed. Another theory, which has found favor with a number of physicists, supposes that the energy is derived from the radio-atoms themselves and is released in consequence of their disintegration. The latter theory involves the conception that the atoms of the radio-elements contain a great store of latent energy, which only manifests itself when the atom breaks up. There is no direct evidence in support of the view that the energy of the radio-elements is derived from external sources, while there is much indirect evidence against it. Some of this evidence will now be considered. There is now no doubt that the α and β rays consist of particles projected with great speed. In order for the α particle to acquire the velocity with which it is expelled, it can be calculated that it would be necessary for it to move freely between two points differing in potential by about five million volts. It is very difficult to imagine any mechanism which could suddenly impress such an enormous velocity on one of the parts of an atom. It seems much more reasonable to suppose that the α and β particles were originally in rapid motion in the atom, and, for some reason, escaped from the atomic system with the velocity they possessed at the instant of their release. There is now undeniable evidence that radioactivity is always accompanied by the production of new kinds of active matter. Some sort of chemical theory is thus required to explain the facts, whether the view is taken that the energy is derived from the atom itself or from external sources. The "external" theory of the origin of the energy was initially advanced to explain only the heat emission of radium. We have seen that this is undoubtedly connected with the expulsion of α particles from the different disintegration products of radium, and that the radium itself only supplies one quarter of the total heat emission, the rest being derived from the emanation and its further products. On such a theory it is necessary to suppose that in radium there are a number of different active substances, whose power of absorbing external energy dies away with the time, at different but definite rates. This still leaves the fundamental difficulty of the origin of these radioactive products unexplained. Unless there is some unknown source of energy in the medium which the radioactive bodies are capable of absorbing, it is difficult to imagine whence the energy demanded by the external theory can be derived. It certainly cannot be from the air itself, for radium gives out heat inside an ice calometer. It cannot be any type of rays such as the radioactive bodies emit, for the radioactivity of radium, and consequently its heating effect, are unaltered by hermetically sealing it in a vessel of lead several inches thick. The evidence, as a whole, is strongly against the theory that the energy is borrowed from external sources, and, unless a number of improbable assumptions are made, such a theory is quite inadequate to explain the experimental facts. On the other hand, the disintegration theory, advanced by Rutherford and Soddy, not only offers a satisfactory explanation of the origin of the energy emitted by the radio-elements, but also accounts for the succession of radioactive bodies. On this theory, a definite, small proportion of the atoms of radioactive matter every second becomes unstable and breaks up with explosive violence. In most cases, the explosion is accompanied by the expulsion of an a particle, in a few cases, by only a β particle, and in others by a and β particles together. On this view, there is at any time present in a radioactive body a proportion of the original matter which is unchanged and the products of the part which has undergone change. In the case of a slowly changing substance like radium, this point of view is in agreement with the observed fact that the spectrum of radium remains unchanged with its age.

The expulsion of an a or β particle or both from the atom leaves behind an atom which is lighter than before and which has different chemical and physical properties. This atom in turn becomes unstable and breaks up, and the process, once started, proceeds from stage to stage with a definite and measurable velocity in each case.

The energy radiated is, on this view, obtained at the expense of the internal energy of the radio-atoms themselves. It does not contradict the principle of the conservation of energy, for the internal energy of the products of the changes, when the process has come to an end, is supposed to be diminished by the amount of energy emitted during the changes. This theory supposes that there is a great store of internal energy in the radio-atoms themselves. This is not in disagreement with the modern views of the electronic constitution of matter, which have been so ably developed by J. J. Thomson, Larmor, and Lorentz. A simple calculation shows that the mere concentration of the electric charges, which on the electronic theory are supposed to be contained in an atom, implies a store of energy in the atom so enormous that, in comparison, the large evolution of energy from the radio-elements is quite insignificant.

Since the energy emitted from the radio-elements is for the most part kinetic in form, it is necessary to suppose that the a and β particles were originally in rapid motion in the atoms from which they are projected. The disintegration theory supposes that it is the atoms and not the molecules which break up. Such a view is necessary to explain the independence of the rate of disintegration of radioactive matter, of wide variations of temperature, and of the action of chemical and physical agents at our command. This must be conceded if the term atom is used in the ordinary chemical sense. It is, however, probable that the atoms of the radio-elements are in reality complex aggregates of known or unknown kinds of matter, which break up spontaneously. This aggregate behaves like an atom and cannot be resolved into simpler forms by external chemical or physical agencies. It breaks up, however, spontaneously with an evolution of energy enormous compared with that released in ordinary chemical changes. This question is further considered in section VIII of this paper.

The disintegration theory assumes that a small fraction of the atoms break up in unit time, but no definite explanation is, as yet, forthcoming of the causes which lead to this explosive disruption of the atom. The experimental results are equally in agreement with the view that each atom contains within itself the potentiality of its final disruption, or with the view that the disintegration is precipitated by the action of some external cause that may lead to the disintegration of the atom in the same way that a detonator is necessary to start certain explosions. The energy set free is, however, not derived from the detonator, but from the substance on which it acts. There is another general view which may possibly lead to an explanation of atomic disruption. If the atom is supposed to consist of electrons or charged bodies in rapid motion, it tends to radiate energy in the form of electromagnetic waves. If an atom is to be permanently stable, the parts of the atom must be so arranged that there is no loss of energy by electromagnetic radiation. J. J. Thomson has investigated certain possible arrangements of electrons in an atom which radiate energy extremely slowly, but which ultimately must break up in consequence of the loss of internal energy. According to present views, it is not such a matter of surprise that atoms do break up as that atoms are so stable as they appear to be. This question of the causes of disintegration is fundamental, and no adequate explanation has yet been put forward.

VII. *Radioactive Products*

Rutherford and Soddy showed that the radioactivity was always accompanied by the appearance of new types of active matter which possessed physical and chemical properties distinct from the parent radio-element. The radioactivity of these products is not permanent, but decays according to an exponential law with the time. The activity I_t and at any time t is given by $I_t = I_o e^{-\lambda t}$ where I_o is the initial activity and λ a constant. Each radioactive product has a definite change-constant which distinguishes it from all other products. These products do not arise simultaneously, but in consequence of a succession of changes in the radio-elements; for example, thorium in breaking up gives rise to *Thorium X*, which behaves as a solid substance soluble in ammonia. This in turn breaks up and gives rise to a gaseous product, the thorium emanation. The emanation is again unstable and gives rise to another type of matter which behaves as a solid and is deposited on the surface of the vessel containing the emanation. It was found that the results would be quantitatively explained on the assumption that the activity of any product at any time is the measure of the rate of production of the next product. This is to be expected, since the activity of any substance is proportional to the number of atoms which break up per second; and since each atom in breaking up gives rise to one atom of the next product together with α or β particles or both, the activity of the parent is a measure of the rate of production of the succeeding product.

Of these radioactive products, the radium emanation has been very closely studied on account of its existence in the gaseous state. It has been shown to be produced by radium at a constant rate. The amount of emanation stored up in a given mass of radium reaches a maximum value when the rate of supply of fresh emanation balances the rate of change of the emanation present.

If q be the number of atoms of emanation produced per second by the radium and N the maximum number present when radioactive equilibrium is reached, then $N = \frac{q}{\lambda}$, where λ is the constant of change of the emanation. This relation has been verified experimentally. The emanation is found to diffuse through air like a gas of heavy molecular weight. It is unattacked by chemical reagents, and in that respect resembles the inert gases of the argon family. It condenses at a definite temperature $-150°C$. Its constant of change is unaffected between the limits of temperature of $450°C$ and $-180°C$. Since the emanation changes into a non-volatile type of matter which is deposited on the surface of vessels, it was to be expected that the volume of the emanation should decrease according to the same law, as it lost its activity. These deductions, based on the theory, have been confirmed in a striking manner by the experiments of Ramsay and Soddy. The radium emanation was chemically isolated and found to be a gas which obeyed Boyle's law. The volume of the emanation observed was of the same order as had been predicted before its separation. The volume was found to decrease with the time according to the same law as the emanation lost its activity. Ramsay and Collie found that the emanation had a new and definite spectrum similar in some respects to that of the argon group of gases.

There can thus be no doubt that the emanation is a transition substance with remarkable properties. Chemically it behaves like an inert gas, and has a definite spectrum, and is condensed by cold. But, on the other hand, the gas is not permanent, but disappears, and is changed into other types of matter. It emits during its changes about one million times as much energy as is emitted during any known chemical change.

From the similarity of the behavior of the emanation of thorium and actinium to that of radium, we may safely conclude that these also are new gases which have only a limited life and change into other substances.

The non-volatile products of the radioactive bodies can be dissolved in strong acids and show definite chemical behavior in solution.

They can be partially separated by electrolysis and by suitable chemical methods. They can be volatilized by the action of high temperature, and their differences in this respect can be utilized to effect in many cases a partial separation of successive products. There can be little doubt that each of these radioactive products is a transition substance, possessing, while it lasts, some definite chemical and physical properties which serve to distinguish it from other products and from the parent element.

The radioactive products derived from each radio-element, together with the type of radiation emitted during their disintegration, are shown graphically in Fig. 1.

Radium   Emanation   Rad·A   Rad B   Rad.C   Rad·D   Rad·E

Thorium   Thor.X   Emanation   Thor.A   Thor.B   Thor.C

Uranium   UraniumX   Final Product

Actinium   Actin X   Emanation   Actin.A   Actin·B   Actin C

The radiations from actinium have not so far been examined sufficiently closely to determine the character of the radiation emitted by each product. There is some evidence that a product, actinium X, exists in actinium corresponding to Th X in thorium. It has not, however, been very closely examined.

The question of nomenclature for the successive products is important. The names Ur X, Th X have been retained, and also the term emanation. The emanation from the three radio-elements in each case gives rise to a non-volatile type of matter which is deposited on the surface of the bodies. The matter initially deposited from the radium emanation is called radium A, radium A changes into B, and B into C, and so on. A similar nomenclature is applied to the further products of the emanation of thorium and actinium. This notation is simple and elastic, and is very useful in mathematical discussion of the theory of successive changes. In the following table a list of the products is given, together with the nature of the radiation and the most marked chemical and physical properties of each product. The time T for each of the products to be half transformed is also added.

Radioactive products.	T	Rays.	Some chemical and physical properties.
URANIUM ↓	2.5×10^8 years.	α	Soluble in excess of ammonium carbonate.
Uranium X ↓	22 days.	β, γ	Insoluble in excess of ammonium carbonate.
Final product.			
THORIUM ↓	10^9 years.	α	Insoluble in ammonia.
Thorium X ↓	4 days.	α	Soluble in ammonia.
Emanation ↓	1 min.	α	Inert gas condenses about $-120°C$.
Thorium A ↓	11 hours.	no rays.	Attaches itself to negative electrode, soluble in strong acids.
Thorium B ↓	55 mins.	α, β, γ	Separable from A by electrolysis.
Final product.			
ACTINIUM ↓			
Actinium X ? ↓			
Emanation ↓	3.9 secs.	α	Gaseous product.
Actinium A ↓	41 mins.	no rays.	Attaches itself to negative electrode, soluble in strong acids.
Actinium B ↓	1.5 mins.	α	Separable from A by electrolysis.
Final product.			
RADIUM ↓	1000 years.	α	
Emanation ↓	4 days.	α	Inert gas, condenses $-150°$ C.
Radium A ↓	3 mins.	α	Attaches itself to negative electrode soluble in strong acids.
Radium B ↓	21 mins.	no rays.	Volatile at $500°C$.
Radium C ↓	28 mins.	α, β, γ	Volatile about $1100°C$.
Radium D ↓	about 40 years.	β, γ	Soluble in sulphuric acid.
Radium E ↓	about 1 year.	α	Attaches itself to bismuth plate in solution, volatilizes at $1000°C$.

The changes which occur in the active deposits from the emanation of radium, thorium, and actinium have been difficult to determine on account of their complexity. For example, in the case of radium, the active deposit, obtained as a result of a long exposure to the emanation, contains quantities of radium A, B, and C. The changes occurring in the active deposit of radium have been determined by P. Curie, Danne, and the writer. The value of T for the three successive changes is 3, 21, and 28 minutes respectively. Radium A gives only a rays, B gives out no rays at all, while C gives out a, β, and γ rays. These results have been deduced by the comparison of the change of activity with time, with the mathematical theory of successive changes. The variation of the activity with time depends upon whether the activity is measured by the a, β, or γ rays. The complicated curves are very completely explained on the hypothesis of three successive changes of the character already mentioned.

The activity of a vessel in which the radium emanation has been stored for some time rapidly falls to a very small fraction after the emanation is withdrawn. There, however, always remains a slight residual activity. The writer has recently examined the activity in ·detail. The residual activity at first mainly consists of β rays, and the activity measured by them does not change appreciably during the period of one year. The a ray activity is at first small, but increases uniformly with the time for the first few months that the activity has been examined. These results receive an explanation on the hypothesis that radium C changes into a product D which emits only β rays. D changes into a product E, which emits only a rays. This view has been confirmed by separating the a ray product by dipping a bismuth plate into the solution containing radium D and E. The probable period of these changes can be deduced from observations of the magnitude of the a and β ray activity at any time. It has been deduced that radium D is probably half transformed in forty years, and radium E is half transformed in about one year. The evidence at present obtained points to the conclusion that radium E is the active constituent present in Marckwald's radio-tellurium, and probably also in the polonium of Mme. Curie.

The changes in the active deposit of thorium have been analyzed by the writer, and the corresponding changes in actinium by Miss Brooks.

The occurrence of a "rayless change" in the active deposits from the emanation of radium, thorium, and actinium is of great interest and importance. As these products do not emit either a or β or γ rays, their presence can only be detected by their effect on the amount of the succeeding products. The action of the rayless change is most clearly brought out in the examination of the variation of activity with time of a body exposed for a very short interval in the presence of the emanations of thorium and actinium. Let us consider, for simplicity, the variation of activity with time for thorium. The activity (measured by the a rays) observed at first is very small, but gradually increases with the time, passes through a maximum, and finally decays according to an exponential law with the time falling to half value in 11 hours. The shape of this curve can be completely explained on the assumption of the two successive changes, the second of which alone gives out rays. The matter deposited on the body during the short exposure consists almost entirely of thorium A. Thorium A changes into B and the breaking up of B gives rise to the activity measured.

Let n_0 =number of particles of thorium A deposited on the body during the time of exposure to the emanation.

Let P and Q be the number of particles of thorium A and B respectively at any time after removal.

Let λ_1, λ_2 be the constants of the two changes.

The number of particles of P existing at any time t is given by $P = n_0 e^{-\lambda_1 t}$. If each atom of A in breaking up gives rise to one atom of B, the increase dQ in the number of Q in the time dt is given by the difference between the number of atoms of B supplied by the change in A and the number of B which break up.

Thus, $$\frac{dQ}{dt} = \lambda_1 P - \lambda_2 Q = \lambda_1 n_0 e^{-\lambda_1 t} - \lambda_2 Q.$$

The solution of this equation is of the form $Q = a e^{-\lambda_1 t} + b e^{-\lambda_2 t}$. Since for a very short exposure $Q = 0$

$$a = -b = \frac{\lambda_1}{\lambda_1 - \lambda_2}$$

and $$Q = \frac{n_0}{\lambda_1 - \lambda_2}(c^{-\lambda_2 t} - e^{-\lambda_1 t}).$$

If thorium A does not give out rays, the activity of the body at any time after removal is proportional to Q. Thus the activity at any time t is proportional to $e^{-\lambda_2 t} - e^{-\lambda_1 t}$. Now the experimental curve of variation of activity is found to be accurately expressed by an equation of this form. A very interesting point arises in settling the values of λ_1 and λ_2 corresponding to the two changes. It is seen that the equation is symmetrical in λ_1 and λ_2 and in consequence is unaltered if the values of λ_1 and λ_2 are interchanged. Now the constant of the change is determined by the observation that the activity finally decays to half value in 11 hours. The theoretical and experimental curves are found to coincide if one of the two products is half transformed in 11 hours and the other in 55 minutes. The comparison of the theoretical and experimental curves does not, however, allow us to settle whether the period of change of thorium A is 55 minutes or 11 hours. In order to settle the point, it is necessary to find some means of separating the products thorium A and B from each other. In the case of thorium, this is done by electrolysing a solution of thorium. Pegram obtained an active product which decayed according to an exponential law with the time falling to half value in a little less than 1 hour. This result shows that the radiating product thorium B has the shorter period. In a similar way, by recourse to electrolysis, it has been found that the change of actinium B has a period of 1.5 minutes.

In the case of radium, P. Curie and Danne utilized the difference in volatility of radium B and C in order to fix the period of the changes.

It is very remarkable that the third successive product of radium, thorium, and actinium should not give out rays. It seems probable that these rayless changes are not of so violent a character as the other changes, and consist either of a rearrangement of the components of the atom or of an expulsion of an a or β particle with so slow a velocity that it fails to ionize the gas. The appearance of such changes in radioactive matter suggests the possibility that ordinary matter may also be undergoing slow "rayless changes," for such changes could not have been detected in the radio-elements, unless its succeeding products emitted rays.

It is seen that the changes occurring in radium, thorium, and actinium are of a very analogous character and indicate that each of these bodies has a very similar atomic constitution.

While there can be no doubt that numerous kinds of radioactive matter with distinct chemical and physical properties are produced in the radio-elements; it is very difficult to obtain direct evidence in some cases that the products are successive and not simultaneous. This is the case for products which have either a very slow or very rapid rate of change compared with the other product. For example, it is difficult to show directly that radium B is the product of radium A and not the direct product of the emanation. In the same way, there is no direct evidence that radium C is the parent of radium D. At the same time the successive nature of these products is indicated by indirect evidence.

There can be little doubt that each of the radioactive products is a distinct chemical substance and possesses some distinguishing physical or chemical properties. There still remains a large amount of chemical work to be done to compare and arrange the chemical properties of these products and to see if the successive products follow any definite law of variation. The electrolytic method can in many cases be used to find the position of the product in the electrochemical series. The products which change most rapidly are present in the least quantity in radium and pitchblende. Only the slower changing products like the radium emanation and radium D and E exist in sufficient quantities to be examined by the balance. It is possible that the products radium A, B, and C may be obtained in sufficient quantity to obtain their spectrum.

VIII. *Connection between the a Particles and Helium*

The discovery of Ramsay and Soddy that helium was produced by the radium emanation was one of the greatest interest and importance, and confirmed in a striking manner the disintegration theory of radioactivity, for the possible production of helium from radioactive matter had been predicted on this theory before the experimental evidence was forthcoming. Ramsay and Soddy found that the presence of helium could not be detected in a tube immediately after the introduction of the emanation, but was observed some time afterwards, showing that the helium arose in consequence of a slow change in the emanation itself or in its further products.

The question of the origin of the helium produced by the radium emanation and its connection with the radioactive changes occurring in the emanation is one of the greatest importance. The experimental evidence so far obtained does not suffice to give a definite answer to this question, but suggests the probable explanation. There has been a tendency to assume that the helium is the final disintegration product of the radium emanation, *i. e.*, it is the inactive substance which remains when the succession of radioactive changes in the emanation have come to an end. There is no evidence in support of such a conclusion, while there is much indirect evidence against it. It has been shown that the emanation which breaks up undergoes three fairly rapid transformations; but after these changes have occurred, the residual matter — radium *D* — is still radioactive, and breaks up slowly, being half transformed in probably about forty years. There then occurs a still further change. Taking into account the minute quantity of the radium emanation initially present in the emanation tube, the amount of the final inactive product would be insignificant after the lapse of a few days or even months. It thus does not seem probable that the helium can be the final product of the radioactive changes. In addition, it has been shown that the *a* particle behaves like a body of about the same mass as the helium atom. The expulsion of a few *a* particles from each of the heavy atoms of radium would not diminish the atomic weight of the residue very greatly. The atomic weight of the atoms of radium *D* and *E* is in all probability of the order of two hundred, since the evidence supports the conclusion that each atom expels one *a* particle at each transformation. In order to explain the presence of helium, it is necessary to look to the other inactive products produced during the radioactive changes. The *a* particles expelled from the radioactive product are themselves non-radioactive. The measurement of the ratio $\frac{e}{m}$ shows that they have an apparent mass intermediate between that of the hydrogen and helium atoms. If the *a* particles consist of any known kind of matter they must be either atoms of hydrogen or of helium. The actual value of $\frac{e}{m}$ has not yet been determined with an accuracy sufficient to give a definite answer to the question. On account of the very slight curvature of the path of the *a* particles in a strong magnetic or electric field, an accurate determination of $\frac{e}{m}$ is beset with great difficulties.

The experimental problem is still further complicated by the fact that the a particles escaping from a mass of radium have not all the same velocity, and in consequence it is difficult to draw a definite conclusion from the observed deviation of the complex pencil of rays.

The results so far obtained are not inconsistent with the view that the a particles are helium atoms, and indeed it is difficult to escape from such a conclusion. On such a view, the helium, which is gradually produced in the emanation tube, is due to the collection of a particles expelled during the disintegration of the emanation and its further products. This conclusion is supported by evidence of another character. It is known that thorium minerals like monazite sand contain a large quantity of helium. In this respect, they do not differ from uranium minerals which are rich in radium. The only common product of the different radioactive substances is the a particle, and the occurrence of helium in all radioactive minerals is most simply explained on the supposition that the a particle is a projected helium atom. This conclusion could be indirectly tested by examining whether helium is produced in other substances besides radium, for example, in actinium and polonium.

The experimental determination of the origin of helium is beset with difficulty on all sides. If the a particle is a helium atom, the total volume of helium produced in an emanation tube should be three times the initial volume of the emanation present, since the emanation in its rapid changes gives rise to three products each of which emits a particles. This is based on the assumption, which seems to be borne out by the experiments, that each atom of each product in breaking up expels one a particle. This at first sight offers a simple experimental means of settling the question, but a difficulty arises in accurately determining the volume of helium produced by a known quantity of the radium emanation. It would be expected that, if the emanation were isolated in a tube and left to stand, the volume of gas in the tube should increase with time in consequence of the liberation of helium. In one case, however, Ramsay and Soddy observed an exactly opposite result. The volume diminished with time to a small fraction of its original value. This diminution of volume was due to the decomposition of the emanation into a non-gaseous type of matter deposited on the walls of the tube, and followed the law of decrease to be expected in such a case, namely, the volume decreased according to an exponential law with the time, falling to half value in four days. The helium produced by the emanation must have been absorbed by the walls of the tube. Such a result is to be expected if the particle is a helium atom, for the a particle is projected with a velocity sufficient to bury itself in the glass to a depth of about $\frac{1}{100}$ mm. This buried helium would probably be in part released by the heating of the tube, such as occurs with the strong electric discharge employed in the spectroscopic detection of helium. Ramsay and Soddy have examined the glass tubes in which the emanation had been confined for some time, to see if the buried helium was released by heat. In some cases, traces of helium were observed.

Accurate measurements of the value of $\frac{e}{m}$ for the a particle, and also an accurate determination of the relative volume of the emanation and the helium produced by it, would probably definitely settle this fundamental question.

Certain very important consequences follow on the assumption that the a particle is, in all cases, an atom of helium. It has already been shown that the radio-elements are transformed into a succession of new substances, most of which in breaking up emit an a particle. On such a view, the atom of radium, thorium, uranium, and actinium must be supposed to be built up in part of helium atoms. In radium, at least five products of the change emit a particles, so that the radium atom must contain at least five atoms of helium. In a similar way, the atoms of actinium and thorium (or if thorium itself be not radioactive, the atom of the active substance present in it) must be compounds of helium. These compounds of helium are not stable, but spontaneously break up into a succession of substances, with an evolution of helium, the disintegration taking place at a definite but different rate at each stage. Such compounds are sharply distinguished in their behavior from the molecular compounds known to chemistry. In the first place, the radioactive compounds disintegrate spontaneously and at a rate that is independent of the physical or chemical forces at our control. Changes of temperature, which exert such a marked influence in altering the rate of molecular reactions are here almost entirely without influence. But the most striking feature of the disintegration is the expulsion, in most cases, of a product of the change with very great velocity — a result never observed in ordinary chemical reactions. This entails an enormous liberation of energy during the change, the amount, in most cases, being about one million times as great as that observed in any known chemical reaction. In order to account for the expulsion of an a or β particle with the observed velocities, it is necessary to suppose that their particles exist in a state of rapid motion in the system from which they escape. Variation of temperature, in most cases, does not seem to affect the stability of the system.

It is well established that the property of radioactivity is inherent in the radio-atoms, since the activity of any radioactive compound depends only on the amount of the element present and is not affected by chemical treatment. As far as observation has gone, both uranium and radium behave as elements in the usual accepted chemical sense.

They spontaneously break up, but the rate of their disintegration seems to be, in most cases, quite independent of chemical control. In this respect, the radioactive bodies occupy a unique position. It seems reasonable to suppose that while the radioactive substances behave chemically as elements, they are, in reality, compounds of simpler kinds of matter, held together by much stronger forces than those which exist between the components of ordinary molecular compounds. Apart from the property of radioactivity, the radio-elements do not show any chemical properties to distinguish them from the non-radioactive elements, except their very high atomic weight. The above considerations thus evidently suggest that the heavier inactive elements may also prove to be composite.

IX. *Origin of the Radio-Elements*

We have seen that the radio-elements are continuously breaking up and giving rise to a succession of new substances. In the case of uranium and thorium, the disintegration proceeds at such a slow rate that in all probability a period of about 1000 million years would be required before half the matter present is transformed. In the case of radium, however, where the process of disintegration proceeds at over one million times the rate of uranium and thorium, it is to be expected that a measurable proportion of the radium should be transformed in a single year. A quantity of radium left to itself must gradually disappear as such in consequence of its gradual transformation into other substances. This conclusion necessarily follows from the known experimental facts. The radium is continuously being transformed into the emanation which in turn is changed into other types of matter. Since there is no evidence that the process is reversible, all the radium present must, in the course of time, be transformed into emanation. The rate at which radium is being transformed can be approximately calculated either from the number of a particles expelled per second or from the observed volume of the emanation produced per second. Both methods of calculation agree in fixing that in a gram of radium about one milligram is transformed per year. From analogy with other radioactive changes, it is to be expected that the rate of change of radium would be always proportional to the amount present. The amount of radium would thus decrease exponentially with the time, falling to half value in about 1000 years. On this point of view, radium behaves in a similar way to the other known products, the only difference being that its rate of change is slower. We have already seen that, in all probability, the product radium D is half transformed in about 40 years and radium E in about one year. In regard to their rate of change, the two substances radium D and E, which are half transformed in about 40 years and 1 year respectively, occupy an intermediate position between the rapidly changing substances like radium A, B, and C and the slowly changing parent substance radium.

If the earth were supposed to have been initially composed of pure radium, the activity 20,000 years later would not be greater than the activity observed in pitchblende to-day. Since there is no doubt that the earth is much older than this, in order to account for the existence of radium at all in the earth, it is necessary to suppose that radium is continuously produced from some other substance or substances. On this view, the present supply of radium represents a condition of approximate equilibrium where the rate of production of fresh radium balances the rate of transformation of the radium already present. In looking for a possible source of radium, it is natural to look to the substances which are always found associated with radium in pitchblende. Uranium and thorium both fulfill the conditions necessary to be a source of radium, for both are found associated with radium and both have a rate of change slow compared with radium. At the present time, uranium seems the most probable source of radium. The activity observed in a good specimen of pitchblende is about what is to be expected, if uranium breaks up into radium. If uranium is the parent of radium, it is to be expected that the amount of radium present in different varieties of pitchblende obtained from different sources should always be proportional to the amount of uranium contained in the minerals. The recent experiments of Boltwood, Strutt, and McCoy indicate that this is very approximately the case. It is not to be expected that the relation should, in all cases, be very exact, since it is not improbable, in some cases, that a portion of the active material may be removed from the mineral, by the action of percolating water or other chemical agencies. The results so far obtained strongly support the view that radium is a product of the disintegration of uranium. It should be possible to obtain direct evidence on this question by examining whether radium appears in uranium compounds which have been initially freed from radium. On account of the delicacy of the electric test of radium by means of its emanation, the question can very readily be put to experimental trial. This has been done for uranium by Soddy and for thorium by the writer, but the results, so far obtained, are negative in character, although, if radium were produced at the rate to be expected from theory, it should very readily have been detected. Such experiments, however, taken over a period of a few months, are not decisive, for it is by no means improbable that the parent element may pass through several slow changes, possibly of a "rayless" character, before it is transformed into radium. In such a case, if these intermediate products are removed by the same chemical process from the parent element, there may be a long period of apparent retardation before the radium appears. The considerations advanced to account for radium apply equally well to actinium, which in all probability, when isolated, will prove to be an element of the same order of activity as radium. The most important problem at present in the study of radioactive minerals is not the attempt to discover and isolate new radioactive substances, but to correlate those already discovered. Some progress has already been made in reducing the number of different radioactive substances and in indicating the origin of some of them. For example, there is no doubt that the " emanating substance " of Giesel contains the same radioactive substance as the actinium of Debierne. In a similar way, there is very strong evidence that the active constituent in the polonium of Mme. Curie is identical with that in the radio-tellurium of Marckwald. The writer has recently shown that the active constituent in radio-tellurium or polonium is, in all probability, a disintegration product of radium (radium E). The same considerations apply to the radio-lead of Hofmann, which is probably identical with the product radium D. It still remains to be shown whether or no there is any direct family connection between the radioactive substances uranium, thorium, radium, and actinium. It seems probable that some at least of these substances will prove to be lineal descendants of a single parent element, in the same way that the radium products are lineal descendants of radium. The subject is capable of direct attack by a combination of physical and chemical methods, and there is every probability that a fairly definite answer will soon be forthcoming.

X. Radioactivity of the Earth and Atmosphere

It is now well established, notably by the work of Elster and Geitel, that radioactive matter is widely distributed both in the earth's crust and atmosphere. There is undoubted evidence of the presence of the radium emanation in the atmosphere, in spring water, and in air sucked up through the soil. It still remains to be settled whether the observed radioactivity of the earth's crust is due entirely to slight traces of the known radioactive elements or to new kinds of radioactive matter. It is not improbable that a close examination of the radioactivity of the different soils may lead to the discovery of radioactive substances which are not found in pitchblende or other radioactive minerals. The extraordinary delicacy of the electroscopic test of radioactivity renders it possible not only to detect the presence in inactive matter of extremely minute traces of a radioactive substance, but also in many cases to settle quickly whether the radioactivity is due to one of the known radio-elements.

The observations of Elster and Geitel render it probable that the radioactivity observed in the atmosphere is due to the presence of radioactive emanations or gases, which are carried to the surface by the escape of underground water and by diffusion through the soil. Indeed, it is difficult to avoid such a conclusion, since there is no evidence that any of the known constituents of the atmosphere are radioactive. Concurrently with observations of the radioactivity of the atmosphere, experiments have been made on the amount of ionization in the atmosphere itself. It is important to settle what part of this ionization is due to the presence of radioactive matter in the atmosphere. Comparisons of the relative amount of active matter and of the ionization in the atmosphere over land and sea will probably throw light on this important problem.

The wide distribution of radioactive matter in the soils which have so far been examined has raised the question whether the presence of radium and other radioactive matter in the earth may not, in part at least, be responsible for the internal heat of the earth. It can readily be calculated that the presence of radium (or equivalent amounts of other kinds of radioactive matter) to the extent of about five parts in one hundred million million by mass would supply as much heat to the earth as is lost at present by conduction to its surface. It is certainly significant that, as far as observation has gone, the amount of radioactive matter present in the soil is of this order of magnitude.

The production of helium from radium indirectly suggests a method of calculation of the age of the deposits of radioactive minerals. It seems reasonable to suppose that the helium always found associated with radioactive minerals is a product of the decomposition of the radioactive matter present. About half of the helium is removed by heating the mineral and the other half by solution. It thus does not seem likely that much of the helium found in the mineral escapes from it, so that the amount present represents the quantity produced since its formation. If the rate of the production of helium by radium (or other radioactive substance) is known, the age of the mineral can at once be estimated from the observed volume of helium stored in the mineral and the amount of radium present. All these factors have, however, not yet been determined with sufficient accuracy to make at present more than a rough estimate of the age of any particular mineral. An estimate of the rate of production of helium by radium has been made by Ramsay and Soddy by an indirect method. It can be deduced from their result that 1 gram of radium produces per year a volume of helium of about 25 cubic mms. at standard pressure and temperature. They, however, consider this to be an underestimate. On the other hand, if the a particle is a helium atom, it can readily be calculated that 1 gram of radium produces per year about 200 cubic mms. of helium.

Let us consider for example the mineral fergusonite. Ramsay and

186 PHYSICS OF THE ELECTRON

Travers have shown that it yields about 1.8 cc. of helium per gra.n and contains about 7 per cent of uranium. It can readily be deduced from known data that each gram of the mineral contains about one four-millionth of a gram of radium. Supposing that one gram of radium produces $\frac{1}{5}$ cc. of helium per year, the age of the mineral is readily seen to be about 40 million years. If the above rate of production of helium by radium is an overestimate, the time will be correspondingly longer. I think there is little doubt that when the data required are accurately known this method can be applied, with considerable confidence, to determine the age of the radioactive minerals.

XI. *Radioactivity of Ordinary Matter*

The property of radioactivity is exhibited to the most marked extent by the radioactive substances found in pitchblende, but it is natural to ask the question whether ordinary matter possesses this property to an appreciable degree. The experiments that have so far been made show conclusively that ordinary matter, if it possesses this property at all, does so to a minute extent compared with uranium. It has been found that all the matter that has so far been examined shows undoubted traces of radioactivity, but it is very difficult to show that the radioactivity observed is not due to a minute trace of known radioactive matter. Even with the extraordinarily delicate methods of detection of radioactivity, the effects observed are so minute that a definite settlement of the question is experimentally very difficult. J. J. Thomson has recently given an account at the Meeting of the British Association of the work done on this subject in the Cavendish Laboratory, and has brought forward experimental evidence that strongly supports the view that ordinary matter does show specific radioactivity. Different substances were found to give out radiations that differed in quality as well as in quantity. A promising beginning has already been made, but a great deal of work still remains to be done before such an important conclusion can be considered to be definitely established.

SHORT PAPER

PROFESSOR R. A. MILLIKAN, of the University of Chicago, presented a paper to this Section on "The Relation between the Radioactivity and the Uranium Content of Certain Minerals," of which the following is an abstract:

In March, 1904, the author, assisted by Mr. H. A. Nichols, Assistant Curator of Geology at the Field Columbian Museum (Chicago), began an investigation of the relation between the radioactivity and the uranium content of uranium-bearing minerals with a view to ascertaining whether the radioactive substances found in pitchblende are not all decomposition products of uranium. If such be the case the ratio between the uranium content and the radioactivity of uranium ought obviously to be constant, in case the assumption may be made that the active products of the decomposition are not washed out of the mineral by percolating water or other agencies.

Since the beginning of this investigation some preliminary results have been published in *Nature* by Boltwood which indicate a constancy in this ratio in the case of a few American ores which he has examined. McCoy (*cf. Ber. d. Chem. Ges.* 36, 3043) has also found a similar indication of constancy in the case of the six different kinds of uranium minerals which he has studied.

The present investigation is not yet complete, but so far as it has gone it furnishes additional evidence in support of the view that uranium is the parent of radium, for it extends somewhat the number of minerals for which the ratio between the activity and the uranium content is approximately constant. The following table gives the results thus far obtained.

Name of mineral	Locality	Per cent of uranium contained	Activity in terms of uranium oxide	% uranium divided by activity	% of departure from mean
Pitchblende	Colorado	59.1	3.24	18.2	4.2
Clevite	Norway	69.3	4.03	17.2	9.4
Gummite	North Carolina	55.4	2.56	21.6	13.6
Pitchblende	Cornwall, Eng.	9.23	.55	16.9	11.6
Autunite	Cornwall, Eng.	6.9	.33	20.8	9.5
Autunite	Saxony	4.0	.205	19.5	3.6
			Mean	19.0	

It will be seen that the departures from the mean value of the ratio amount in some cases to as much as 13 %, but this was found to be no more than the differences which might be obtained by "resurfacing" the same specimen of a given substance.

The measurements on activity were all made as follows: three hundred mg. of the very carefully powdered mineral were spread as uniformly as possible over three square inches of a metal sheet. This sheet was then placed upon the lower plate of an air-condenser which was connected with one pair of quadrants of an electrometer, the other pair being earthed. The condenser-plates were ten cm. on a side and 3 cm. apart. A potential of one hundred and thirty volts was applied to the upper condenser-plate, and the rate of charging of the electrometer noted. The potential to which the needle of the electrometer was charged was one hundred and twenty-five volts. The chemical analyses were all made by Mr. Nichols.

BIBLIOGRAPHY: DEPARTMENT OF PHYSICS

(Prepared for the Department by the courtesy of Professor Henry Crew)

GENERAL PHYSICS

ARRHENIUS, Lehrbuch der kosmischen Physik.
GUILLAUME and POINCARÉ, Rapports présentés au Congrès International de physique. (Paris, 1900.)
POYNTING and THOMSON, Text-Book of Physics.
WATSON, Text-Book of Physics (Longmans).
WINKELMANN, Handbuch der Physik, 6 vols. 2d ed.
VIOLLE, Cours de Physique.

DYNAMICS

CLIFFORD, Elements of Dynamic.
HERTZ, Die Principien der Mechanik.
LAGRANGE, Mécanique Analytique.
LAMB, Hydrodynamics.
LOVE, Treatise on the Mathematical Theory of Elasticity.
MINCHEN, Treatise on Statics.
NEWTON, Principia.
THOMSON and TAIT, Treatise on Natural Philosophy.
WEBSTER, Dynamics.
WIEN, Lehrbuch der Hydrodynamik.

SOUND

BLASERNA, Theory of Sound.
DONKIN, Acoustics.
HELMHOLTZ, Sensations of Tone. (Trans. by Ellis.)
RAYLEIGH, Theory of Sound.

HEAT

BOLTZMANN, Vorlesungen über Gastheorie.
EWING, Steam Engine and other Heat Engines.
FOURIER, Analytical Theory of Heat. (Trans. by Freeman.)
GUILLAUME, Thermométrie.
JEANS, Dynamical Theory of Gases.
MAXWELL, Theory of Heat.
MEYER, Kinetic Theory of Gases. (Trans. by Baynes.)
PLANCK, Thermodynamics. (Trans. by Ogg.)
PRESTON, Theory of Heat.

LIGHT

CZAPSKI, Theorie der Optischen Instrumente.
DRUDE, Theory of Optics. (Trans. by Mann and Millikan.)
EDSER, Light for Students.
KAYSER, Handbuch der Spectroscopie.
KELVIN, Baltimore Lectures.
LARMOR, Ether and Matter.

Ludwig Boltzmann
1844–1906

SECTION C — APPLIED MATHEMATICS

(*Hall 7, September 24, 3 p. m.*)

CHAIRMAN: PROFESSOR ARTHUR G. WEBSTER, Clark University, Worcester, Mass.
SPEAKERS: PROFESSOR LUDWIG BOLTZMANN, University of Vienna.
PROFESSOR HENRI POINCARÉ, The Sorbonne; Member of the Institute of France.
SECRETARY: PROFESSOR HENRY T. EDDY, University of Minnesota.

THE RELATIONS OF APPLIED MATHEMATICS

BY LUDWIG BOLTZMANN

(*Translated from the German by Professor S. Epsteen, University of Chicago*)

[Ludwig Boltzmann, Professor of Physics, University of Vienna, since 1902. b. Vienna, Austria, 1840. Studied, Vienna, Heidelberg, and Berlin. Professor of Physico-Mathematics, University of Gratz, 1869–73; Professor of Mathematics, University of Vienna, 1873–76; Professor of Experimental Physics, University of Gratz, 1876–90; Professor of Theoretical Physics, University of Munich, 1891–95; *ibid.* University of Vienna, 1895–1900; Professor of Physics, University of Leipzig, 1900–02. **Author of** *Vorlesungen über Maxwell's Theorie der Elekt rizitat und des Lichts; Vorlesungen über Kinetische Gastheorie; Vorlesungen über die Prinzipe der Mechanik.*]

MY present lecture has been put under the heading of applied mathematics, while my activity as a teacher and investigator belongs to the science of physics. The immense gap which divides the latter science into two distinct camps has almost nowhere been so sharply emphasized as in the division of the lecture material of this scientific congress, which covers such an enormous range of subjects that one may designate it as a flood, or, to preserve local coloring, as a Niagara of scientific lectures. I speak of the division of physics into theoretical and experimental. Although I have been assigned, as representative of theoretical physics, to "A. — Normative Science," experimental physics appears much later under " C.—Physical Science." Between them lie history, science of language, literature, art, and science of religion. Over all this, however, the theoretical physicist must extend his hand to the experimental physicist. We shall therefore not be able to avoid entirely the question of the justification of dividing physics into two parts and, in particular, into theoretical and experimental.

Let us listen first of all to an investigator of a time when natural science had not yet grown beyond its first beginnings, to Emmanuel Kant. Kant requires of each science that it should be developed logically from unified principles and firmly established theories. Natural science seems to him a primary science only in so far as it rests on a mathematical basis. Thus, he does not reckon the chemistry of his day among the sciences, because it rests merely upon an empirical basis and lacks a unified, regulative principle.

From this point of view theoretical physics is preferred to experimental physics, and occupies, in a sense, a higher rank. Experimental physics was merely to gather the material, but it remained for the theoretical physics to form the structure.

But the succession in the order of rank becomes reversed when we take into account the acquisitions of the last decades as well as the progress which is to be expected in the immediate future. The chain of experimental discoveries of the last century received a fitting completion with the discovery of the Röntgen rays. Connected with these there appear in the present century a multitude of new rays, with the most enigmatical properties, which have the profoundest effects upon our conceptions of nature. The more enigmatical these newly discovered facts are, and the more they seem at first to contradict our present conceptions, the greater the successes which they promise for the future. But this is not the occasion for the discussion of these experimental triumphs. I must leave to the representatives of experimental physics at this Congress the prolific problem of portraying all of the fruits which have hitherto been gathered in this domain, one might almost say, daily, and those which are to be expected.

The representative of theoretical physics scarcely finds himself in an equally fortunate position. Great activity does indeed prevail in this domain. One could almost say that it is in process of revolution. Only how much less tangible are the results here attained in comparison with those in experimental physics! It appears here that in a certain sense experimentation deserves precedence over all theory. An immediate fact is at once comprehensible. Its fruits may become evident in the shortest time, such as the various applications of the Röntgen rays and the utilization of the Hertz waves in wireless telegraphy. The battle which the theories have to fight is, however, an infinitely wearisome one; indeed, it seems as if certain disputed questions which existed from the beginning will live as long as the science.

Every firmly established fact remains forever unchangeable; at most, it may be generalized, completed, additions may be made, but it cannot be completely upset. Thus it is explained why the development of experimental physics is continuously progressive, never making a sudden jump, and never visited by great tremblings and revolutions. It occurs only in rare instances that something which was regarded as a fact turns out afterwards to have been an error, and in such cases the explanations of the errors follow soon, and they are not of great influence on the structure of the science as a whole.

It is, indeed, strongly emphasized that every established and logically recognized truth must remain incontrovertible. Although this cannot be doubted, experience teaches that the structure of our theories is by no means composed entirely of such incontrovertibly established truths. They are composed rather of many arbitrary pictures of the connections between phenomena, of so-called hypotheses.

Without some departure, however slight, from direct observation, a theory or even an intelligibly connected practical description for predicting the facts of nature cannot exist. This is equally true of the old theories whose foundations have become questionable, and of the most modern ones, which are resigning themselves to a great illusion if they regard themselves as free from hypotheses.

The hypotheses may perhaps be indefinite, or may be in the shape of mathematical formulae, or the thought may be equivalent to the latter, but expressed in words. In the latter cases the agreement with given data may be checked step by step; a complete revolution of that previously constructed is indeed not absolutely impossible, as, for example, if the law of the conservation of energy should turn out to be incorrect. But such a revolution will be exceedingly rare and highly improbable.

Such an indefinite, slightly specialized theory might serve as a guiding thread for experiments whose purpose is a detailed development of knowledge previously acquired and which is proceeding in barren channels; beyond this its usefulness does not reach.

In contradistinction to these are the hypotheses which give the imagination room for play and by boldly going beyond the material at hand afford continual inspiration for new experiments, and are thus pathfinders for the most unexpected discoveries. Such a theory will indeed be subject to change, a very complicated mass of information will be brought together and will then be replaced by a new and more comprehensive theory in which the old one will be the picture of a limited type of phenomena. Examples of this are the theory of emission in regard to the description of the phenomena of catoptrics and dioptrics, the hypothesis of an elastic ether in the representation of the phenomena of interference and refraction of light, and the notion of the electric fluid in the description of the phenomena of electrostatics.

Moreover the theories which proudly designate themselves as free from hypotheses are not exempt from great revolutions; thus, no one will doubt that the so-called theory of energy will have completely to alter its form if it desires to remain effective.

The accusation has been made that physical hypotheses have sometimes proved injurious and have delayed the progress of the science. This accusation is based chiefly upon the rôle which the hypothesis of the electric fluid has played in the development of the theory of electricity. This hypothesis was brought to a high stage of perfection by Wilhelm Weber, and the general recognition which his works found in Germany did indeed stand in the road of the theory of Maxwell. In a similar manner Newton's emanation theory stood in the way of the theory of undulations. But such inconveniences can scarcely be entirely avoided in the future. It will always be the tendency to complete as far as possible the prevailing view, and to make it self-sufficient whenever such a theory is self-consistent and does not in any way lead to a contradiction, whether it consist of mechanical models, of geometrical pictures, or of mathematical formulas. It will always be possible that a new theory will arise which has not yet been tested by experiment and which will represent a much larger field of phenomena. In such cases the older theory will count the largest following until this field of phenomena is brought into the range of experiment, and decisive tests demonstrate the superiority of the newer one. It is certainly useful, if the theory of Weber be always held up as a warning example, that one should bear in mind the essential progressiveness of the intellect. The services of Weber are not decreased by this, however; Maxwell himself speaks of his theory with the greatest wonder. Indeed, this instance cannot be taken into consideration against the usefulness of hypotheses, since Maxwell's theory contained as much of the hypothetical as any other. And this was eliminated only after it became generally known through Hertz, Poynting, and others.

The accusation has also been raised against hypotheses in physics that the creation and development of mathematical methods for the computation of the hypothetical molecular motions has been useless and even harmful. This accusation I cannot recognize as substantiated. Were it so, the theme selected for my present lecture would be an unfortunate one; and this fact may excuse me for having lingered on this much-discussed subject and for having sought to justify the use of hypotheses in physics.

I have not chosen for the thesis of my present lecture the entire development of physical theory. Several years ago I treated this subject at the German *Naturforscherversammlung* in Munich, and although some new developments have taken place since then, I should have to repeat myself a great deal. Moreover, one who has committed himself to one faction is not in a position to judge the other factions in a completely objective manner. I do not refer to a criticism of its value; my lecture shall not criticise, but shall judge. I am also convinced of the value of the views of my opponents and only arise to repel them when they attempt to belittle mine. But one can scarcely give as complete an account according to subject-matter, and an exposition of the inter-relations of all ideas in the views of another, as in his own.

I shall therefore select as the goal of my lecture to-day not merely the kinetic theory of molecules, but, moreover, a highly specialized branch of it. Far from denying that it contains hypotheses, I must rather characterize it as a bold advance beyond the facts of observation. And I nevertheless do not consider it unworthy of this occasion; this much faith do I have in hypotheses which present certain peculiarities of observation in a new light or which bring forth relations among them which cannot be reached by other methods. We must indeed be mindful of the fact that hypotheses require and are capable of continuous development, and are only then to be abandoned when all the relations which they represent can be better understood in some other manner.

To the above-mentioned problems, which are as old as the science and still unsolved, belongs the one if matter is continuous, or if it is to be considered as made up of discrete parts, of very many, but not in the mathematical sense infinite, individuals. This is one of the difficult questions which form the boundary of philosophy and physics.

Even some decades ago, scientists felt very shy of going deeply into the discussion of such questions. The one before us is too real to be entirely avoided; but one cannot discuss it without touching on some profounder still, such as upon the nature of the law of causation, of matter, of force, and so forth. The latter are the ones of which it was said that they did not trouble the scientist, that they belonged entirely to philosophy. To-day the situation is different, there is evident a tendency among scientists to consider philosophic questions, and properly so. One of the first rules of science is never to trust blindly to the instrument with which one works, but to test it in all directions. How, then, are we to trust blindly to inherited and historically developed conceptions, particularly when there are instances known where they led us into error? But in the examination of even the simplest elements, where is the boundary between science and philosophy at which we should pause?

I hope that none of the philosophers present will take offense or perceive an accusation, if I say boldly that by assigning this question to philosophy the resulting success has been rather meagre. Philosophy has done noticeably little toward the explanation of these questions, and from her own one-sided point of view she can do so just as little as natural science can from hers. If real progress is possible, it is only to be expected by coöperation of both of these sciences. May I therefore be pardoned if I touch slightly upon these questions

271

although not a specialist; their connection with the aim of my lecture is very intimate.

Let us consult the famous thinker already quoted, Emmanuel Kant, on the question if matter is continuous, or if it is composed of atoms. He treats of this in his *Antimonies*. Of all the questions there raised, he shows that both the *pro* and *con* can be logically demonstrated. It can be shown rigorously that there is no limit to the divisibility of matter while an infinite divisibility contradicts the laws of logic. Kant shows likewise that a beginning and end of time, a boundary where space ceases, are as inconceivable as absolutely endless duration, absolutely endless extension.

This is by no means the sole instance where philosophical thought becomes tangled in contradictions; indeed, one finds them at every step. The ordinary things of philosophy are sources of insolvable riddles; to explain our perceptions it invents the concept of matter and then finds that it is altogether unsuited to possess perception itself or to generate perception in a spirit. With consummate acumen it constructs the concept of space, or of time, and finds that it is absolutely impossible that things should exist in this space, that events should occur during this time. It finds insurmountable difficulties in the relation of cause to effect, of body and soul, in the possibility of consciousness, in short, everywhere and in everything. Indeed, it finally finds it inexplicable and self-contradictory that anything can exist at all, that something originated and is capable of continuing, that everything can be classified according to our categories, nor that there is a quite perfect classification. Such a classification will always be a variable one and adapted to the requirements of the moment. Also the breaking up of physics into theoretical and experimental is merely a consequence of the prevalent division of methods and will not last forever.

My present thesis is quite different from the one that certain questions are beyond the boundary of human comprehension. For according to the latter, there is a deficiency, an incompleteness in the human intelligence, while I consider the existence of these questions, these problems, as an illusion. By superficial consideration it seems astonishing, after this illusion is recognized, that the impulse to answer those questions does not cease. Habit of thought is much too powerful to release us.

It is here as with the ordinary illusion which continues operative after its cause is recognized. In consequence of this is the feeling of uncertainty, of want of satisfaction which the scientist feels when he philosophizes. These illusions will yield but very slowly and gradually, and I consider it as one of the chief problems of philosophy to set forth clearly the uselessness of reaching beyond the limits of our habits of thought and to strive, in the choice and combination of concepts and words, to give the most useful expression of facts in a manner which is independent of our inherited habits. Then all these complications and contradictions must vanish. It must be made clear what is stone in the structure of our thoughts and what is mortar, and the oppressive sentiment, that the simplest things are the most inexplicable and the most trivial are the most mysterious, becomes mere imagination-change.

To call upon logic seems to me as if one were to put on for a trip into the mountains a long flowing robe, which always wrapped itself about the feet so that one fell at the first steps while on the level. The source of this kind of logic is the immoderate trust in the so-called laws of thought. It is certain that we could not gather experience did we not have certain forms of connecting phenomena, that is to say, of thought, innate. If we wish to call these laws of thought, they are indeed *a priori* to the extent that they accompany every experience in our soul, or if we prefer, in our brain. Only nothing seems to me less reasonable than the conclusion from the reasoning in this sense to certainty, to infallibility. These laws of thought have been developed according to the same laws of evolution as the optical apparatus of the eye, the acoustic apparatus of the ear, and the pumping arrangements of the heart. In the course of human development everything useless was eliminated, and thus a unity and finish arose which might be mistaken for infallibility. Thus the perfection of the eye, of the ear, of the arrangement of the heart excite our admiration, without the absolute perfection of these organs being emphasized, however. Just so little should the laws of thought be regarded as absolutely infallible. They are the very ones which have developed with regard to seizing that which is most necessary and practically useful in the maintenance of life. With these, the results of experimental investigation show more relation than the examination of the mechanism of thought. We should, therefore, not be surprised that the customary forms of thought for the abstract are not entirely suited to practical applications in far removed problems of philosophy, and that they have not become applicable since the days of Thales. Therefore the simplest things seem to be the most puzzling to the philosopher. And he finds everywhere contradictions; these are nothing more, however, than useless, incorrect facsimiles of that which is given us through our thoughts. In facts there can be no contradictions. As soon as contradictions seem unavoidable we must test, extend, and seek to modify that which we call laws of thought, but which are only inherited, customary representations, preserved for aeons, for the description of practical needs. Just as to the inherited discoveries of the cylinder, the carriage, the plow, numerous artificial ones have been consciously added, so must we improve, artificially and con-

273

sciously, our likewise inherited concepts. Our problem cannot be to quote facts before the judgment seat of our laws of thought, but to fit our mental representations and concepts to the facts. Since we attempt to express with clearness such complicated processes merely by words, written, spoken, or inwardly thought, it might also be said that we should combine the words in such wise as to give the most appropriate expression of the facts, that the relations indicated by our words should be most adequate for the relations among the actualities. When the problem is enunciated in this fashion, its appropriate solution may still offer the greatest difficulties, but one knows then the end in view and will not stumble on self-made difficulties.

Much that is useless in the usage and in the bearing of the nature of life is brought forth by a method of treatment which, being useful in most cases, becomes through habit a second nature, until one cannot set it aside when it becomes inapplicable somewhere. I say that the adaptability goes beyond the point aimed at. This happens frequently in the commonplaces of thought, and becomes the source of apparent contradictions between the laws of thought and the world, as well as between the laws of thought themselves.

Thus, the regularity of the phenomena of nature is the fundamental condition for all cognition; thus comes the habit of inquiring of everything the cause, the non-resisting compulsion, and we inquire also concerning the cause, why everything must have a cause. In fact people strove for a long time to determine if cause and effect is a necessary bond or merely an accidental sequence, and if it did or did not have a unique meaning to say that a certain particular phenomenon was connected with, and a necessary consequence of, a definite group of other phenomena.

Similarly, something is said to be useful, valuable, if it satisfies the needs of the individual or of humanity; but we go beyond the mark if we ask concerning the value of life itself, if such it seem to have, because it has no purpose outside of itself. The same happens when we strive vainly to explain the simplest concepts, out of which all others are built, by means of simpler ones still, to explain the simplest fundamental laws.

We should not attempt to deduce nature from our concepts, but should adapt the latter to the former. We should not believe our inherited rules of thought to be conditions preceding our more complicated experiences, for they are not so for the simplest essentials. They arose slowly in connection with simple experiences and passed, by heredity, to the more highly organized being. Thus is explained how synthetic judgments arise which were formed by our ancestors and were born in us, and are in this sense *a priori*. Their great power is also seen in this way, but not their infallibility.

In saying that such judgments as "everything is red or is not red" are results of experience, I do not mean that every person checks this empty truth by experience, but that he learns that his parents called everything either red or not red and that he preserves this nomenclature.

It might seem as if we had gone somewhat deeply into philosophical questions, but I believe that the views we have reached could not have been attained in a shorter and simpler manner. For we have reached an impartial judgment how the question of the atomistic structure of matter is to be viewed. We shall not invoke the law of thought that there is no limit to the divisibility of matter. This law is of no more value than if a naïve person were to say that no matter where he went upon the earth the plumb-line directions seemed always to be parallel and therefore there were no antipodes.

On the one hand we shall start from facts only, and on the other we shall take nothing into consideration except the effort to attain to the most adequate expression of these facts.

Regarding the first point, the numerous facts of the theory of heat, of chemistry, of crystallography, show that bodies which are apparently continuous do not by any means fill the entire volume indistinguishably and uniformly with matter. Indeed, it appears that the space which they occupy is filled with innumerably many individuals, molecules, and atoms, which are extraordinarily small, but not infinitely small in the mathematical sense. Their sizes can be computed in different manners and always with the same result.

The fruitfulness of this line of thought has been verified in the most recent time. All the phenomena which are observed with the cathode rays, the Becquerel rays, etc., indicate that we are dealing with diminutive, moving particles, electrons. After a vigorous battle, this view vanquished completely the opposing explanation of these phenomena by the theory of undulations. Not only did the former theory give a better explanation of the previously known facts, it inspired new experiments and permitted the prediction of unknown phenomena, and thus it developed into an atomistic theory of electricity. If it continue to develop with the same success as in past years, if phenomena, such as the one observed by Ramsay on the transmutation of radium into helium, do not remain isolated, this theory promises deductions concerning the nature and structure of atoms as yet undreamed of. Computation shows that electrons are much smaller than the atoms of ponderable matter; and the hypothesis that the atoms are built up of many elements, as well as various interesting views on the character and structure of this composition, is to-day on every tongue. The word atom should not lead us into error, it comes from a past time; no physicist ascribes indivisibility to the atoms.

It is not my intention to confine the thought merely to the above facts and their resulting consequences; these are not sufficient to carry through the question as to the finite or infinite divisibility of matter. If we are going to think of the atoms of chemistry as made up of electrons, what would hinder us from considering the electrons as particles filled with rarefied, continuous matter? We shall adhere faithfully to the previously developed philosophical principles and shall examine in the most unhampered manner the concepts themselves in order to express them in a consistent and most useful form.

It appears now, that we are unable to define the infinite in any other way except as the limit of continually increasing magnitudes, at least no one has hitherto been able to set up any other intelligible conception of the infinite. Should we desire a verbal picture of the continuum, we must first think of a large finite number of particles which are endowed with certain properties and study the totality of these particles. Certain properties of this totality may approach a definite limit as the number of particles is increased, and their size decreased. It can be asserted, concerning these properties, that they belong to the continuum, and it is my opinion that this is the only self-consistent definition of a continuum which is endowed with certain properties.

The question if matter is composed of atoms or is continuous becomes then the question if the observed properties are accurately satisfied by the assumption of an exceedingly great number of such particles or, by increasing number, their limit. We have not indeed answered the old philosophical question, but we are cured of the effort to answer it in a senseless and hopeless manner. The thought-process, that we must investigate the properties of a finite totality and then let the number of members of this totality increase greatly, remains the same in both cases. It is nothing other than the abbreviated expression in algebraic symbols of exactly the same thought when, as often happens, differential equations are made the basis of a mathematical-physical theory.

The members of the totality which we select as the picture of the material body cannot be thought of as absolutely at rest, for there would then be no motion of any kind, nor can the members be thought of as relatively at rest in one and the same body, for in this case it would be impossible to account for the fluids. No effort has been made to subject them to anything more than to the general laws of mechanics. In order to explain nature we shall therefore select a totality of an exceedingly large number of very minute fundamental individuals which are constantly in motion, and which are subject to the laws of mechanics. But an objection is raised that will be an appropriate introduction to the final considerations of this lecture. The fundamental equations of mechanics do not alter their form in the slightest way when the algebraic sign of the time is changed. All pure mechanical events can therefore occur equally well in one sense as in its opposite, that is, in the sense of increasing time or of diminishing time. We remark, however, that in ordinary life future and past do not coincide as completely as the directions right and left, but that the two are distinctly different.

This becomes still more definite by means of the second law of the mechanical theory of heat, which asserts that when an arbitrary system of bodies is left to itself, uninfluenced by other bodies, the sense in which changes of condition occur can be assigned. A certain function of the condition of all the bodies, the entropy, can be determined, which is such that every change that occurs must be in the sense of carrying with it an increase of this function; thus, with increasing time the entropy increases. This law is indeed an abstraction, just as the principle of Galileo; for it is impossible, in strict rigor, to isolate a system of bodies from all others. But since it has given correct results hitherto, in connection with all the other laws, we assume it to be correct, just as in the case of the principle of Galileo.

It follows from this law that every closed system of bodies must tend toward a definite final condition for which the entropy is a maximum. The outcome of this law, that the universe must come to a final state in which nothing more can occur, has caused astonishment; but this outcome is only comprehensible on the assumption that the universe is finite and subject to the second law of the mechanical theory of heat. If one regards the universe as infinite, the above-mentioned difficulties of thought arise again if one does not consider the infinite as a mere limit of the finite. Since there is nothing analogous to the second law in the differential equations of mechanics, it follows that it can be represented mechanically only by the initial conditions. In order to find the assumptions suitable for this purpose, we must reflect that, to explain the apparent continuity of bodies, we had to assume that every family of atoms, or more generally, of mechanical individuals, existed in incredibly many different initial positions. In order to treat this assumption mathematically, a new science was founded whose problem is, not the study of the motion of a single mechanical system, but of the properties of complexes of very many mechanical systems which begin with a great variety of initial conditions. The task of systematizing this science, of compiling it into a large book, and of giving it a characteristic name, was executed by one of the greatest American scholars, and in regard to abstract thinking, purely theoretic investigation, perhaps the greatest, Willard Gibbs, the recently deceased professor at Yale University. He called this science statistical mechanics, and it falls naturally into two parts. The first investigates the conditions under which the outwardly visible properties of a complex of very many mechanical individuals is not in any wise altered; this first part I shall call statistical statics. The second part investigates the gradual changes of these outwardly visible properties when those conditions are not fulfilled; it may be called statistical dynamics. At this point we may allude to the broad view which is opened by applying this science to the statistics of animated beings, of human society, of sociology, etc., and not merely upon mechanical particles. A development of the details of this science would only be possible in a series of lectures and by means of mathematical formulas. Apart from mathematical difficulties it is not free from difficulties of principle. It is based upon the theory of probabilities. The latter is as exact as any other branch of mathematics if the concept of equal probabilities, which cannot be deduced from the other fundamental notions, is assumed. It is here as in the method of least squares which is only free from objection when certain definite assumptions are made concerning the equal probability of elementary errors. The existence of this fundamental difficulty explains why the simplest result of statistical statics, the proof of Maxwell's speed law among the molecules of a gas, is still being disputed.

The theorems of statistical mechanics are rigorous consequences of the assumptions and will always remain valid, just as all well-founded mathematical theorems. But its application to the events of nature is the prototype of a physical hypothesis. Starting from the simplest fundamental assumption of the equal probabilities, we find that aggregates of very many individuals behave quite analogously as experience shows of the material world. Progressive or visible rotary motion must always go over into invisible motion of the minutest particles, into heat, as Helmholtz characteristically says: ordered motion tends always to go over into not ordered motion; the mixture of different substances as well as of different temperatures, the points of greater or less intense molecular motion, must always tend toward homogeneity. That this mixture was not complete from the start, that the universe began in such an improbable state, belongs to the fundamental hypotheses of the entire theory; and it may be said that the reason for this is as little known as the reason why the universe is just so and not otherwise. But we may take a different point of view. Conditions of great mixture and great differences in temperature are not absolutely impossible according to the theory but are very highly improbable. If the universe be considered as large enough there will be, according to the laws of probability, here and there places of the size of fixed stars, of altogether improbable distributions. The development of such a spot would be one-sided both in its structure and subsequent dissolution. Were there thinking beings at such a spot their impressions of time would be the same as ours, although the course of events in the universe as a whole would not be one-sided. The above-developed theory does indeed go boldly beyond our experience, but it has the merit which every such theory should have of showing us the facts of experience in an entirely new light and of inspiring us to new thought and reflection. In contradistinction to the first fundamental law, the second one is merely based on probability, as Gibbs pointed out in the '70's of the last century.

I have not avoided philosophical questions, in the firm hope that coöperation between philosophy and natural science will give new sustenance to both; indeed, that only in this manner a consistent argument can be carried through. I agree with Schiller when he says to the scientists and philosophers of his day, " Let there be strife between you, and the union will come speedily;" I believe that the time for this union has now arrived.

Henri Poincaré
1854–1912

THE PRINCIPLES OF MATHEMATICAL PHYSICS

BY JULES HENRI POINCARÉ

(Translated from the French by George Bruce Halsted, Kenyon College)

[Jules Henri Poincaré, Professor University of Paris, and the Polytechnic School; Member of Bureau of Longitude. b. Nancy, April 29, 1854. D.Sc. August 3, 1879; D.Sc. Cambridge and Oxford, 1879; Charge of the Course of the Faculty of Sciences at Caen; Master of Conference of the Faculty of Sciences of Paris, 1881; Professor of the same Faculty, 1886; Member of the Institute of France, 1887; Corresponding Member of the National Academy of Washington; Philosophical Society of Philadelphia; the Academies of Berlin, London, St. Petersburg, Vienna, Rome, Munich, Göttingen, Bologna, Turin, Naples, Venice, Amsterdam, Copenhagen, Stockholm, etc. Written books and numerous articles for reviews and periodicals.]

WHAT is the actual state of mathematical physics? What are the problems it is led to set itself? What is its future? Is its orientation on the point of modifying itself?

Will the aim and the methods of this science appear in ten years to our immediate successors in the same light as to ourselves; or, on the contrary, are we about to witness a profound transformation? Such are the questions we are forced to raise in entering to-day upon our investigation.

If it is easy to propound them, to answer is difficult.

If we feel ourselves tempted to risk a prognostication, we have, to resist this temptation, only to think of all the stupidities the most eminent savants of a hundred years ago would have uttered, if one had asked them what the science of the nineteenth century would be. They would have believed themselves bold in their predictions, yet after the event how very timid we should have found them.

Mathematical physics, we know, was born of celestial mechanics, which engendered it at the end of the eighteenth century, at the moment when the latter was attaining its complete development. During its first years especially, the infant resembled in a striking way its mother.

The astronomic universe is formed of masses, very great without doubt, but separated by intervals so immense that they appear to us only as material points. These points attract each other in the inverse ratio of the square of the distances, and this attraction is the sole force which influences their movements. But if our senses were sufficiently subtle to show us all the details of the bodies which the physicist studies, the spectacle we should there discover would scarcely differ from what the astronomer contemplates. There also we should see material points, separated one from another by intervals enormous in relation to their dimensions, and describing orbits following regular laws.

These infinitesimal stars are the atoms. Like the stars properly so called, they attract or repel each other, and this attraction or this repulsion directed following the straight line which joins them, depends only on the distance. The law according to which this force varies as function of the distance is perhaps not the law of Newton, but it is an analogous law; in place of the exponent — 2, we have probably a different exponent, and it is from this change of exponent that springs all the diversity of physical phenomena, the variety of qualities and of sensations, all the world colored and sonorous which surrounds us, — in a word, all nature.

Such is the primitive conception in all its purity. It only remains to seek in the different cases what value should be given to this exponent in order to explain all the facts. It is on this model that Laplace, for example, constructed his beautiful theory of capillarity; he regards it only as a particular case of attraction, or as he says of universal gravitation, and no one is astonished to find it in the middle of one of the five volumes of the *Mécanique céleste*.

More recently Briot believed he had penetrated the final secret of optics in demonstrating that the atoms of ether attract each other in the inverse ratio of the sixth power of the distance; and does not Maxwell himself say somewhere that the atoms of gases repel each other in the inverse ratio of the fifth power of the distance? We have the exponent — 6, or — 5 in place of the exponent — 2, but it is always an exponent.

Among the theories of this period, one alone is an exception, that of Fourier; in it are indeed atoms, acting at a distance one upon the other; they mutually transmit heat, but they do not attract, they never budge. From this point of view, the theory of Fourier must have appeared to the eyes of his contemporaries, even to Fourier himself, as imperfect and provisional.

This conception was not without grandeur; it was seductive, and many among us have not finally renounced it; we know that we shall attain the ultimate elements of things only by patiently disentangling the complicated skein that our senses give us; that it is necessary to advance step by step, neglecting no intermediary; that our fathers were wrong in wishing to skip stations; but we believe that when we shall have arrived at these ultimate elements, there again will be found the majestic simplicity of celestial mechanics.

Neither has this conception been useless; it has rendered us an inestimable service, since it has contributed to make precise in us the fundamental notion of the physical law.

I will explain myself; how did the ancients understand law? It was for them an internal harmony, static, so to say, and immutable;

or it was like a model that nature constrained herself to imitate. A law for us is not that at all; it is a constant relation between the phenomenon of to-day and that of to-morrow; in a word, it is a differential equation.

The ideal form of physical law is the law of Newton which first covered it; and then how has one to adapt this form to physics? by copying as much as possible this law of Newton, that is, in imitating celestial mechanics.

Nevertheless, a day arrived when the conception of central forces no longer appeared sufficient, and this is the first of those crises of which I just now spoke.

Then investigators gave up trying to penetrate into the detail of the structure of the universe, to isolate the pieces of this vast mechanism, to analyze one by one the forces which put them in motion, and were content to take as guides certain general principles which have precisely for their object the sparing us this minute study.

How so? Suppose that we have before us any machine; the initial wheel-work and the final wheel-work alone are visible, but the transmission, the intermediary wheels by which the movement is communicated from one to the other are hidden in the interior and escape our view; we do not know whether the communication is made by gearing or by belts, by connecting-rods or by other dispositives.

Do we say that it is impossible for us to understand anything about this machine so long as we are not permitted to take it to pieces? You know well we do not, and that the principle of the conservation of energy suffices to determine for us the most interesting point. We easily ascertain that the final wheel turns ten times less quickly than the initial wheel, since these two wheels are visible; we are able thence to conclude that a couple applied to the one will be balanced by a couple ten times greater applied to the other. For that there is no need to penetrate the mechanism of this equilibrium and to know how the forces compensate each other in the interior of the machine; it suffices to be assured that this compensation cannot fail to occur.

Well, in regard to the universe, the principle of the conservation of energy is able to render us the same service. This is also a machine, much more complicated than all those of industry, and of which almost all the parts are profoundly hidden from us; but in observing the movement of those that we can see, we are able, by aid of this principle, to draw conclusions which remain true whatever may be the details of the invisible mechanism which animates them.

The principle of the conservation of energy, or the principle of Mayer, is certainly the most important, but it is not the only one;

there are others from which we are able to draw the same advantage. These are:

The principle of Carnot, or the principle of the degradation of energy.

The principle of Newton, or the principle of the equality of action and reaction.

The principle of relativity, according to which the laws of physical phenomena should be the same, whether for an observer fixed, or for an observer carried along in a uniform movement of translation; so that we have not and could not have any means of discerning whether or not we are carried along in such a motion.

The principle of the conservation of mass, or principle of Lavoisier.

I would add the principle of least action.

The application of these five or six general principles to the different physical phenomena is sufficient for our learning of them what we could reasonably hope to know of them.

The most remarkable example of this new mathematical physics is, beyond contradiction, Maxwell's electro-magnetic theory of light.

We know nothing of the ether, how its molecules are disposed, whether they attract or repel each other; but we know that this medium transmits at the same time the optical perturbations and the electrical perturbations; we know that this transmission should be made conformably to the general principles of mechanics, and that suffices us for the establishment of the equations of the electromagnetic field.

These principles are results of experiments boldly generalized; but they seem to derive from their generality itself an eminent degree of certitude.

In fact the more general they are. the more frequently one has the occasion to check them, and the verifications, in multiplying themselves, in taking forms the most varied and the most unexpected, finish by no longer leaving place for doubt.

Such is the second phase of the history of mathematical physics, and we have not yet emerged from it.

Do we say that the first has been useless? that during fifty years science went the wrong way, and that there is nothing left but to forget so many accumulated efforts as vicious conceptions condemned in advance to non-success?

Not the least in the world; the second phase could not have come into existence without the first?

The hypothesis of central forces contained all the principles; it involved them as necessary consequences; it involved both the conservation of energy and that of masses, and the equality of action and reaction; and the law of least action, which would appear, it is true, not as experimental verities, but as theorems, and of which the enunciation would have at the same time a something more precise and less general than under their actual form.

It is the mathematical physics of our fathers which has familiarized us little by little with these divers principles; which has taught us to recognize them under the different vestments in which they disguise themselves. One has to compare them to the data of experience, to find how it was necessary to modify their enunciation so as to adapt them to these data; and by these processes they have been enlarged and consolidated.

So we have been led to regard them as experimental verities; the conception of central forces became then a useless support, or rather an embarrassment, since it made the principles partake of its hypothetical character.

The frames have not therefore broken, because they were elastic; but they have enlarged; our fathers, who established them, did not work in vain, and we recognize in the science of to-day the general traits of the sketch which they traced.

Are we about to enter now upon the eve of a second crisis? Are these principles on which we have built all about to crumble away in their turn? For some time, this may well have been asked.

In hearing me speak thus, you think without doubt of radium, that grand revolutionist of the present time, and in fact I will come back to it presently; but there is something else.

It is not alone the conservation of energy which is in question; all the other principles are equally in danger, as we shall see in passing them successively in review.

Let us commence with the principle of Carnot. This is the only one which does not present itself as an immediate consequence of the hypothesis of central forces; more than that, it seems, if not directly to contradict that hypothesis, at least not to be reconciled with it without a certain effort.

If physical phenomena were due exclusively to the movements of atoms whose mutual attraction depended only on the distance, it seems that all these phenomena should be reversible; if all the initial velocities were reversed, these atoms, always subjected to the same forces, ought to go over their trajectories in the contrary sense, just as the earth would describe in the retrograde sense this same elliptic orbit which it describes in the direct sense, if the initial conditions of its movement had been reversed. On this account, if a physical phenomenon is possible, the inverse phenomenon should be equally so, and one should be able to reascend the course of time.

But it is not so in nature, and this is precisely what the principle of Carnot teaches us; heat can pass from the warm body to the cold body; it is impossible afterwards to make it reascend the inverse way and reëstablish differences of temperature which have been effaced.

Motion can be wholly dissipated and transformed into heat by friction; the contrary transformation can never be made except in a partial manner.

We have striven to reconcile this apparent contradiction. If the world tends toward uniformity, this is not because its ultimate parts, at first unlike, tend to become less and less different, it is because, shifting at hazard, they end by blending. For an eye which should distinguish all the elements, the variety would remain always as great, each grain of this dust preserves its originality and does not model itself on its neighbors; but as the blend becomes more and more intimate, our gross senses perceive no more than the uniformity. Behold why, for example, temperatures tend to a level, without the possibility of turning backwards.

A drop of wine falls into a glass of water; whatever may be the law of the internal movements of the liquid, we soon see it colored to a uniform rosy tint, and from this moment, however well we may shake the vase, the wine and the water do not seem capable of further separation. Observe what would be the type of the reversible physical phenomenon: to hide a grain of barley in a cup of wheat is easy; afterwards to find it again and get it out is practically impossible.

All this Maxwell and Boltzmann have explained; the one who has seen it most clearly, in a book too little read because it is a little difficult to read, is Gibbs, in his *Elementary Principles of Statistical Mechanics.*

For those who take this point of view, the principle of Carnot is only an imperfect principle, a sort of concession to the infirmity of our senses; it is because our eyes are too gross that we do not distinguish the elements of the blend; it is because our hands are too gross that we cannot force them to separate; the imaginary demon of Maxwell, who is able to sort the molecules one by one, could well constrain the world to return backward. Can it return of itself? That is not impossible; that is only infinitely improbable.

The chances are that we should long await the concourse of circumstances which would permit a retrogradation, but soon or late they would be realized, after years whose number it would take millions of figures to write.

These reservations, however, all remained theoretic and were not very disquieting, and the principle of Carnot retained all its practical value.

But here the scene changes.

The biologist, armed with his microscope, long ago noticed in his preparations disorderly movements of little particles in suspension: this is the Brownian movement; he first thought this was a vital phenomenon, but he soon saw that the inanimate bodies danced with no less ardor than the others; then he turned the matter over to the physicists. Unhappily, the physicists remained long uninterested in this question; the light is focused to illuminate the microscopic preparation, thought they; with light goes heat; hence inequalities of temperature and interior currents produce the movements in the liquid of which we speak.

M. Gouy, however, looked more closely, and he saw, or thought he saw, that this explanation is untenable, that the movements become more brisk as the particles are smaller, but that they are not influenced by the mode of illumination.

If, then, these movements never cease, or rather are reborn without ceasing, without borrowing anything from an external source of energy, what ought we to believe? To be sure, we should not renounce our belief in the conservation of energy, but we see under our eyes now motion transformed into heat by friction, now heat changed inversely into motion, and that without loss since the movement lasts forever. This is the contrary of the principle of Carnot.

If this be so, to see the world return backward, we no longer have need of the infinitely subtle eye of Maxwell's demon; our microscope suffices us. Bodies too large, those, for example, which are a tenth of a millimeter, are hit from all sides by moving atoms, but they do not budge, because these shocks are very numerous and the law of chance makes them compensate each other: but the smaller particles receive too few shocks for this compensation to take place with certainty and are incessantly knocked about. And thus already one of our principles is in peril.

We come to the principle of relativity: this not only is confirmed by daily experience, not only is it a necessary consequence of the hypothesis of central forces, but it is imposed in an irresistible way upon our good sense, and yet it also is battered.

Consider two electrified bodies; though they seem to us at rest, they are both carried along by the motion of the earth; an electric charge in motion, Rowland has taught us, is equivalent to a current; these two charged bodies are, therefore, equivalent to two parallel currents of the same sense and these two currents should attract each other. In measuring this attraction, we measure the velocity of the earth; not its velocity in relation to the sun or the fixed stars, but its absolute velocity.

I know it will be said that it is not its absolute velocity that is measured, but its velocity in relation to the ether. How unsatisfactory that is! Is it not evident that from a principle so understood we could no longer get anything? It could no longer tell us anything just because it would no longer fear any contradiction.

If we succeed in measuring anything, we should always be free to say that this is not the absolute velocity in relation to the ether, it might always be the velocity in relation to some new unknown fluid with which we might fill space.

Indeed, experience has taken on itself to ruin this interpretation of the principle of relativity; all attempts to measure the velocity of the earth in relation to the ether have led to negative results. This time experimental physics has been more faithful to the principle than mathematical physics; the theorists, to put in accord their other general views, would not have spared it; but experiment has been stubborn in confirming it.

The means have been varied in a thousand ways and finally Michelson has pushed precision to its last limits; nothing has come of it. It is precisely to explain this obstinacy that the mathematicians are forced to-day to employ all their ingenuity.

Their task was not easy, and if Lorentz has gotten through it, it is only by accumulating hypotheses.

The most ingenious idea has been that of local time.

Imagine two observers who wish to adjust their watches by optical signals; they exchange signals, but as they know that the transmission of light is not instantaneous, they take care to cross them.

When the station B perceives the signal from the station A, its clock should not mark the same hour as that of the station A at the moment of sending the signal, but this hour augmented by a constant representing the duration of the transmission. Suppose, for example, that the station A sends its signal when its clock marks the hour 0, and that the station B perceives it when its clock marks the hour t. The clocks are adjusted if the slowness equal to t represents the duration of the transmission, and to verify it the station B sends in its turn a signal when its clock marks 0; then the station A should perceive it when its clock marks t. The time-pieces are then adjusted. And in fact, they mark the same hour at the same physical instant, but on one condition, namely, that the two stations are fixed. In the contrary case the duration of the transmission will not be the same in the two senses, since the station A, for example, moves forward to meet the optical perturbation emanating from B, while the station B flies away before the perturbation emanating from A. The watches adjusted in that manner do not mark, therefore, the true time; they mark what one may call the *local time*, so that one of them goes slow on the other. It matters little, since we have no means of perceiving it. All the phenomena which happen at A, for example, will be late, but all will be equally so, and the observer who ascertains them will not perceive it, since his watch is slow; so, as the principle of relativity would have it, he will have no means of knowing whether he is at rest or in absolute motion.

Unhappily, that does not suffice, and complementary hypotheses are necessary; it is necessary to admit that bodies in motion undergo a uniform contraction in the sense of the motion. One of the diameters of the earth, for example, is shrunk by $\frac{1}{200,000,000}$ in consequence of the motion of our planet, while the other diameter retains its normal length. Thus, the last little differences find themselves compensated. And then there still is the hypothesis about forces. Forces, whatever be their origin, gravity as well as elasticity, would be reduced in a certain proportion in a world animated by a uniform translation; or, rather, this would happen for the components perpendicular to the translation; the components parallel would not change.

Resume, then, our example of two electrified bodies; these bodies repel each other, but at the same time if all is carried along in a uniform translation, they are equivalent to two parallel currents of the same sense which attract each other. This electro-dynamic attraction diminishes, therefore, the electro-static repulsion, and the total repulsion is more feeble than if the two bodies were at rest. But since to measure this repulsion we must balance it by another force, and all these other forces are reduced in the same proportion, we perceive nothing.

Thus, all is arranged, but are all the doubts dissipated?

What would happen if one could communicate by non-luminous signals whose velocity of propagation differed from that of light? If, after having adjusted the watches by the optical procedure, one wished to verify the adjustment by the aid of these new signals, then would appear divergences which would render evident the common translation of the two stations. And are such signals inconceivable, if we admit with Laplace that universal gravitation is transmitted a million times more rapidly than light?

Thus, the principle of relativity has been valiantly defended in these latter times, but the very energy of the defense proves how serious was the attack.

Let us speak now of the principle of Newton, on the equality of action and reaction.

This is intimately bound up with the preceding, and it seems indeed that the fall of the one would involve that of the other. Thus we should not be astonished to find here the same difficulties.

Electrical phenomena, we think, are due to the displacements of little charged particles, called electrons, immersed in the medium that we call ether. The movements of these electrons produce perturbations in the neighboring ether; these perturbations propagate themselves in every direction with the velocity of light, and in turn other electrons, originally at rest, are made to vibrate when the perturbation reaches the parts of the ether which touch them.

The electrons, therefore, act upon one another, but this action is not direct, it is accomplished through the ether as intermediary.

Under these conditions can there be compensation between action and reaction, at least for an observer who should take account only of the movements of matter, that is to say, of the electrons, and who should be ignorant of those of the ether that he could not see? Evidently not. Even if the compensation should be exact, it could not be simultaneous. The perturbation is propagated with a finite velocity; it, therefore, reaches the second electron only when the first has long ago entered upon its rest.

This second electron, therefore, will undergo, after a delay, the action of the first, but certainly it will not react on this, since around this first electron nothing any longer budges.

The analysis of the facts permits us to be still more precise. Imagine for example, a Hertzian generator, like those employed in wireless telegraphy; it sends out energy in every direction; but we can provide it with a parabolic mirror, as Hertz did with his smallest generators, so as to send all the energy produced in a single direction.

What happens, then, according to the theory? It is that the apparatus recoils as if it were a gun and as if the energy it has projected were a bullet; and that is contrary to the principle of Newton, since our projectile here has no mass, it is not matter, it is energy.

It is still the same, moreover, with a beacon light provided with a reflector, since light is nothing but a perturbation of the electromagnetic field. This beacon light should recoil as if the light it sends out were a projectile. What is the force that this recoil should produce? It is what one has called the Maxwell-Bartholdi pressure. It is very minute, and it has been difficult to put it into evidence even with the most sensitive radiometers; but it suffices that it exists.

If all the energy issuing from our generator falls on a receiver, this will act as if it had received a mechanical shock, which will represent in a sense the compensation of the recoil of the generator; the reaction will be equal to the action, but it will not be simultaneous; the receiver will move on but not at the moment when the generator recoils. If the energy propagates itself indefinitely without encountering a receiver, the compensation will never be made.

Do we say that the space which separates the generator from the receiver and which the perturbation must pass over in going from the one to the other is not void, that it is full not only of ether, but of air; or even in the interplanetary spaces of some fluid subtle but still ponderable; that this matter undergoes the shock like the receiver at the moment when the energy reaches it, and recoils in its turn when the perturbation quits it? That would save the principle of Newton, but that is not true.

If energy in its diffusion remained always attached to some material substratum, then matter in motion would carry along light with it, and Fizeau has demonstrated that it does nothing of the sort, at least for air. This is what Michelson and Morley have since confirmed.

One may suppose also that the movements of matter, properly so called, are exactly compensated by those of the ether; but that would lead us to the same reflections as just now. The principle so extended would explain everything, since whatever might be the visible movements, we should always have the power of imagining hypothetical movements which compensated them.

But if it is able to explain everything, this is because it does not permit us to foresee anything; it does not enable us to decide between different possible hypotheses, since it explains everything beforehand. It therefore becomes useless.

And then the suppositions that it would be necessary to make on the movements of the ether are not very satisfactory.

If the electric charges double, it would be natural to imagine that the velocities of the divers atoms of ether double also, and for the compensation, it would be necessary that the mean velocity of the ether quadruple.

This is why I have long thought that these consequences of theory, contrary to the principle of Newton, would end some day by being abandoned, and yet the recent experiments on the movements of the electrons issuing from radium seem rather to confirm them.

I arrive at the principle of Lavoisier on the conservation of masses: in truth this is one not to be touched without unsettling all mechanics.

And now certain persons believe that it seems true to us only because we consider in mechanics merely moderate velocities, but that it would cease to be true for bodies animated by velocities comparable to that of light. These velocities, it is now believed, have been realized; the cathode rays or those of radium may be formed of very minute particles or of electrons which are displaced with velocities smaller no doubt than that of light, but which might be its one tenth or one third.

These rays can be deflected, whether by an electric field, or by a magnetic field, and we are able by comparing these deflections, to measure at the same time the velocity of the electrons and their mass (or rather the relation of their mass to their charge). But when it was seen that these velocities approached that of light, it was decided that a correction was necessary.

These molecules, being electrified, could not be displaced without agitating the ether; to put them in motion it is necessary to overcome a double inertia, that of the molecule itself and that of the ether. The total or apparent mass that one measures is composed, therefore, of two parts: the real or mechanical mass of the molecule and the electro-dynamic mass representing the inertia of the ether.

The calculations of Abraham and the experiments of Kaufmann have then shown that the mechanical mass, properly so called, is null, and that the mass of the electrons, or, at least, of the negative electrons, is of exclusively electro-dynamic origin. This forces us to change the definition of mass; we cannot any longer distinguish mechanical mass and electro-dynamic mass, since then the first would vanish; there is no mass other than electro-dynamic inertia. But in this case the mass can no longer be constant, it augments with the velocity, and it even depends on the direction, and a body animated by a notable velocity will not oppose the same inertia to the forces which tend to deflect it from its route, as to those which tend to accelerate or to retard its progress.

There is still a resource; the ultimate elements of bodies are electrons, some charged negatively, the others charged positively. The negative electrons have no mass, this is understood; but the positive electrons, from the little we know of them, seem much greater. Perhaps they have, besides their electro-dynamic mass, a true mechanical mass. The veritable mass of a body would, then, be the sum of the mechanical masses of its positive electrons, the negative electrons not counting; mass so defined could still be constant.

Alas, this resource also evades us. Recall what we have said of the principle of relativity and of the efforts made to save it. And it is not merely a principle which it is a question of saving, such are the indubitable results of the experiments of Michelson.

Lorentz has been obliged to suppose that all the forces, whatever be their origin, were affected with a coefficient in a medium animated by a uniform translation; this is not sufficient; it is still necessary, says he, that *the masses of all the particles be influenced by a translation to the same degree as the electro-magnetic masses of the electrons.*

So the mechanical masses will vary in accordance with the same laws as the electro-dynamic masses; they cannot, therefore, be constant.

Need I point out that the fall of the principle of Lavoisier involves that of the principle of Newton? This latter signifies that the centre of gravity of an isolated system moves in a straight line; but if there is no longer a constant mass, there is no longer a centre of gravity, we no longer know even what this is. This is why I said above that the experiments on the cathode rays appeared to justify the doubts of Lorentz on the subject of the principle of Newton.

From all these results, if they are confirmed, would arise an entirely new mechanics, which would be, above all, characterized by this fact, that no velocity could surpass that of light, any more than any temperature could fall below the zero absolute, because bodies would oppose an increasing inertia to the causes, which would tend to accelerate their motion; and this inertia would become infinite when one approached the velocity of light.

Nor for an observer carried along himself in a translation he did not suspect could any apparent velocity surpass that of light; there would then be a contradiction, if we recall that this observer would not use the same clocks as a fixed observer, but, indeed, clocks marking "local time."

Here we are then facing a question I content myself with stating. If there is no longer any mass, what becomes of the law of Newton?

Mass has two aspects, it is at the same time a coefficient of inertia and an attracting mass entering as factor into Newtonian attraction. If the coefficient of inertia is not constant, can the attracting mass be? That is the question.

At least, the principle of the conservation of energy yet remains to us, and this seems more solid. Shall I recall to you how it was in its turn thrown into discredit? This event has made more noise than the preceding and it is in all the records.

From the first works of Becquerel, and, above all, when the Curies had discovered radium, one saw that every radio-active body was an inexhaustible source of radiations. Its activity would seem to subsist without alteration throughout the months and the years. This was already a strain on the principles; these radiations were in fact energy, and from the same morsel of radium this issued and forever issued. But these quantities of energy were too slight to be measured; at least one believed so and was not much disquieted.

The scene changed when Curie bethought himself to put radium into a calorimeter; it was seen then that the quantity of heat incessantly created was very notable.

The explanations proposed were numerous; but in so far as no one of them has prevailed over the others, we cannot be sure there is a good one among them.

Sir William Ramsay has striven to show that radium is in process of transformation, that it contains a store of energy enormous but not inexhaustible.

The transformation of radium, then, would produce a million times more of heat than all known transformations; radium would wear itself out in 1250 years; you see that we are at least certain to be settled on this point some hundreds of years from now. While waiting our doubts remain.

In the midst of so many ruins what remains standing? The principle of least action has hitherto remained intact, and Larmor appears to believe that it will long survive the others; in reality, it is still more vague and more general.

In presence of this general ruin of the principles, what attitude will mathematical physics take?

And first, before too much perplexity, it is proper to ask if all this is really true. All these apparent contradictions to the principles are encountered only among infinitesimals; the microscope is necessary to see the Brownian movement; electrons are very light; radium is very rare, and no one has ever seen more than some milligrams of it at a time.

And, then, it may be asked if, beside the infinitesimal seen, there be not another infinitesimal unseen counterpoise to the first.

So, there is an interlocutory question, and, as it seems, only experiment can solve it. We have, therefore, only to hand over the matter to the experimenters, and, while waiting for them to determine the question finally, not to preoccupy ourselves with these disquieting problems, but quietly continue our work, as if the principles were still uncontested. We have much to do without leaving the domain where they may be applied in all security; we have enough to employ our activity during this period of doubts.

And as to these doubts, is it indeed true that we can do nothing to disembarrass science of them? It may be said, it is not alone experimental physics that has given birth to them; mathematical physics has well contributed. It is the experimenters who have seen radium throw out energy, but it is the theorists who have put in evidence all the difficulties raised by the propagation of light across a medium in motion; but for these it is probable we should not have become conscious of them. Well, then, if they have done their best to put us into this embarrassment, it is proper also that they help us to get out of it.

They must subject to critical examination all these new views I have just outlined before you, and abandon the principles only after having made a loyal effort to save them.

What can they do in this sense? That is what I will try to explain.

Among the most interesting problems of mathematical physics, it is proper to give a special place to those relating to the kinetic theory of gases. Much has already been done in this direction, but much still remains to be done. This theory is an eternal paradox. We have reversibility in the premises and irreversibility in the conclusions; and between the two an abyss. Statistic considerations, the law of great numbers, do they suffice to fill it? Many points still remain obscure to which it is necessary to return, and doubtless many times. In clearing them up, we shall understand better the sense of the principle of Carnot and its place in the *ensemble* of dynamics, and we shall be better armed to interpret properly the curious experiment of Gouy, of which I spoke above.

Should we not also endeavor to obtain a more satisfactory theory of the electro-dynamics of bodies in motion? It is there especially, as I have sufficiently shown above, that difficulties accumulate. Evidently we must heap up hypotheses, we cannot satisfy all the principles at once; heretofore, one has succeeded in safeguarding some only on condition of sacrificing the others; but all hope of obtaining better results is not yet lost. Let us take, therefore, the theory of Lorentz, turn it in all senses, modify it little by little, and perhaps everything will arrange itself.

Thus in place of supposing that bodies in motion undergo a contraction in the sense of the motion, and that this contraction is the same whatever be the nature of these bodies and the forces to which they are otherwise submitted, could we not make an hypothesis more simple and more natural?

We might imagine, for example, that it is the ether which is modified when it is in relative motion in reference to the material medium which it penetrates, that when it is thus modified, it no longer transmits perturbations with the same velocity in every direction. It might transmit more rapidly those which are propagated parallel to the medium, whether in the same sense or in the opposite sense, and less rapidly those which are propagated perpendicularly. The wave surfaces would no longer be spheres, but ellipsoids, and we could dispense with that extraordinary contraction of all bodies.

I cite that only as an example, since the modifications one might essay would be evidently susceptible of infinite variation.

It is possible also that the astronomer may some day furnish us data on this point; he it was in the main who raised the question in making us acquainted with the phenomenon of the aberration of light. If we make crudely the theory of aberration, we reach a very curious result. The apparent positions of the stars differ from their real positions because of the motion of the earth, and as this motion is variable, these apparent positions vary. The real position we cannot know, but we can observe the variations of the apparent position. The observations of the aberration show us, therefore, not the movement of the earth, but the variations of this movement; they cannot, therefore, give us information about the absolute motion of the earth. At least this is true in first approximation, but it would be no longer the same if we could appreciate the thousandths of a second. Then it would be seen that the amplitude of the oscillation depends not alone on the variation of the motion, variation which is well known, since it is the motion of our globe on its elliptic orbit, but on the mean value of this motion; so that the constant of aberration would not be altogether the same for all the stars, and the differences would tell us the absolute motion of the earth in space.

This, then, would be, under another form, the ruin of the principle of relativity. We are far, it is true, from appreciating the thousandths of a second, but after all, say some, the total absolute velocity of the earth may be much greater than its relative velocity with respect to the sun. If, for example, it were 300 kilometers per second in place of 30, this would suffice to make the phenomena observable.

I believe that in reasoning thus we admit a too simple theory of aberration. Michelson has shown us, I have told you, that the physical procedures are powerless to put in evidence absolute motion; I am persuaded that the same will be true of the astronomic procedures, however far one pushes precision.

However that may be, the data astronomy will furnish us in this regard will some day be precious to the physicist. While waiting, I believe the theorists, recalling the experience of Michelson, may anticipate a negative result, and that they would accomplish a useful work in constructing a theory of aberration which would explain this in advance.

But let us come back to the earth. There also we may aid the experimenters. We can, for example, prepare the ground by studying profoundly the dynamics of electrons; not, be it understood, in starting from a single hypothesis, but in multiplying hypotheses as much as possible. It will be, then, for the physicists to utilize our work in seeking the crucial experiment to decide between these different hypotheses.

This dynamics of electrons can be approached from many sides, but among the ways leading thither is one which has been somewhat neglected, and yet this is one of those which promise us most of surprises. It is the movements of the electrons which produce the line of the emission spectra; this is proved by the phenomenon of Zeemann; in an incandescent body, what vibrates is sensitive to the magnet, therefore electrified. This is a very important first point, but no one has gone farther; why are the lines of the spectrum distributed in accordance with a regular law?

These laws have been studied by the experimenters in their least details; they are very precise and relatively simple. The first study of these distributions recalled the harmonics encountered in acoustics; but the difference is great. Not only the numbers of vibrations are not the successive multiples of one number, but we do not even find anything analogous to the roots of those transcendental equations to which so many problems of mathematical physics conduct us: that of the vibrations of an elastic body of any form, that of the Hertzian oscillations in a generator of any form, the problem of Fourier for the cooling of a solid body.

The laws are simpler, but they are of wholly other nature, and to cite only one of these differences, for the harmonics of high order the number of vibrations tends toward a finite limit, instead of increasing indefinitely.

That has not yet been accounted for, and I believe that there we have one of the most important secrets of nature. Lindemann has made a praiseworthy attempt, but, to my mind, without success; this attempt should be renewed. Thus we shall penetrate, so to say, into the inmost recess of matter. And from the particular point of view which we to-day occupy, when we know why the vibrations of incandescent bodies differ from ordinary elastic vibrations, why the electrons do not behave themselves like the matter which is familiar to us, we shall better comprehend the dynamics of electrons and it will be perhaps more easy for us to reconcile it with the principles.

Suppose, now, that all these efforts fail, and after all I do not believe they will, what must be done? Will it be necessary to seek to mend the broken principles in giving what we French call a *coup de pouce* ? That is evidently always possible, and I retract nothing I have formerly said.

Have you not written, you might say if you wished to seek a quarrel with me, have you not written that the principles, though of experimental origin, are now unassailable by experiment because they have become conventions? And now you have just told us the most recent conquests of experiment put these principles in danger. Well, formerly I was right and to-day I am not wrong.

Formerly I was right, and what is now happening is a new proof of it. Take, for example, the calorimeter experiment of Curie on radium. Is it possible to reconcile that with the principle of the conservation of energy?

It has been attempted in many ways; but there is among them one I should like you to notice.

It has been conjectured that radium was only an intermediary, that it only stored radiations of unknown nature which flashed through space in every direction, traversing all bodies, save radium, without being altered by this passage and without exercising any action upon them. Radium alone took from them a little of their energy and afterward gave it out to us in divers forms.

What an advantageous explanation, and how convenient! First, it is unverifiable and thus irrefutable. Then again it will serve to account for any derogation whatever to the principle of Mayer; it responds in advance not only to the objection of Curie, but to all the objections that future experimenters might accumulate. This new and unknown energy would serve for everything. This is just what I have said, and we are thereby shown that our principle is unassailable by experiment.

And after all, what have we gained by this *coup de pouce* ?

The principle is intact, but thenceforth of what use is it?

It permitted us to foresee that in such or such circumstance we could count on such a total quantity of energy; it limited us; but now where there is put at our disposition this indefinite provision of new energy, we are limited by nothing; and as I have written elsewhere, if a principle ceases to be fecund, experiment, without contradicting it directly, will be likely to condemn it.

This, therefore, is not what would have to be done, it would be necessary to rebuild anew.

If we were cornered down to this necessity, we should moreover console ourselves. It would not be necessary to conclude that science can weave only a Penelope's web, that it can build only ephemeral constructions, which it is soon forced to demolish from top to bottom with its own hands.

As I have said, we have already passed through a like crisis. I have shown you that in the second mathematical physics, that of the principles, we find traces of the first, that of the central forces; it will be just the same if we must learn a third.

When an animal exuviates, and breaks its too narrow carapace to make itself a fresh one, we easily recognize under the new envelope the essential traits of the organism which have existed.

We cannot foresee in what way we are about to expand; perhaps it is the kinetic theory of gases which is about to undergo development and serve as model to the others. Then, the facts which first appeared to us as simple, thereafter will be merely results of a very great number of elementary facts which only the laws of chance make coöperate for a common end. Physical law will then take an entirely new aspect; it will no longer be solely a differential equation, it will take the character of a statistical law.

Perhaps, likewise, we should construct a whole new mechanics, of which we only succeed in catching a glimpse, where inertia increasing with the velocity, the velocity of light would become an impassable limit.

The ordinary mechanics, more simple, would remain a first approximation, since it would be true for velocities not too great, so that we should still find the old dynamics under the new.

We should not have to regret having believed in the principles, and even, since velocities too great for the old formulas would always be only exceptional, the surest way in practice would be still to act as if we continued to believe in them. They are so useful, it would be necessary to keep a place for them. To determine to exclude them altogether would be to deprive one's self of a precious weapon. I hasten to say in conclusion we are not yet there, and as yet nothing proves that the principles will not come forth from the combat victorious and intact.

SHORT PAPERS

Three short papers were read in the Section of Applied Mathematics, the first by Professor Henry T. Eddy, of the University of Minnesota, on "The Electromagnetic Theory and the Velocity of Light."

The second paper was presented by Professor Alexander Macfarlane, of Chatham, Ontario, "On the Exponential Notation in Vector-analysis."

The third paper was presented by Professor James McMahon, of Cornell University, "On the Use of N-fold Riemann Spaces in Applied Mathematics."

PICTURE CREDITS

Administration Building, Washington University, Headquarters of the International Congress of Arts and Science—Reprinted with permission: *Popular Science Monthly,* Vol. 66, No. 1, November 1904, p. 98

Festival Hall, in which the opening exercises of the International Congress of Arts and Science were held—Reprinted from "The Universal Exposition of 1904" by David R. Francis, Louisiana Purchase Exposition Company, St. Louis, MO, 1913, p. 289

Carl Barus (1856–1935), AIP Niels Bohr Library

Ludwig Boltzmann (1844–1906), AIP Niels Bohr Library, Landé Collection

Dewitt Bristol Brace (1859–1905), *Physical Review,* Vol. 24 (1905)

Benno Erdman (1851–1921), Reprinted with permission: *Popular Science Monthly,* Vol. 66, No. 1, November 1904, p. 19

Paul Langevin (1872–1946), copyright by Fred Stein

Arthur Lalanne Kimball (1856–1922), Amherst College Archives

Simon Newcomb (1835–1909), AIP Niels Bohr Library, E. Scott Barr Collection

Edward L. Nichols (1854–1937), Cornell University Archives

Francis Eugene Nipher (1847–1926), AIP Niels Bohr Library

Wilhelm Ostwald (1853–1932), Reprinted with permission: *Popular Science Monthly,* Vol. 66, No. 1, November 1904, p. 20

Henri Poincaré (1854–1912), AIP Niels Bohr Library

Ernest Rutherford (1871–1937), Cambridge University Library, Rutherford Collection; courtesy AIP Niels Bohr Library

Robert Simpson Woodward (1848–1924), AIP Niels Bohr Library